ENVIRONMENTAL RESEARCH ADVANCES

FOREST CONSERVATION

METHODS, MANAGEMENT AND CHALLENGES

ENVIRONMENTAL RESEARCH ADVANCES

Additional books and e-books in this series can be found on Nova's website under the Series tab.

ENVIRONMENTAL RESEARCH ADVANCES

FOREST CONSERVATION

METHODS, MANAGEMENT AND CHALLENGES

PEDRO V. EISENLOHR
EDITOR

Copyright © 2019 by Nova Science Publishers, Inc.

All rights reserved. No part of this book may be reproduced, stored in a retrieval system or transmitted in any form or by any means: electronic, electrostatic, magnetic, tape, mechanical photocopying, recording or otherwise without the written permission of the Publisher.

We have partnered with Copyright Clearance Center to make it easy for you to obtain permissions to reuse content from this publication. Simply navigate to this publication's page on Nova's website and locate the "Get Permission" button below the title description. This button is linked directly to the title's permission page on copyright.com. Alternatively, you can visit copyright.com and search by title, ISBN, or ISSN.

For further questions about using the service on copyright.com, please contact:
Copyright Clearance Center
Phone: +1-(978) 750-8400   Fax: +1-(978) 750-4470   E-mail: info@copyright.com.

NOTICE TO THE READER

The Publisher has taken reasonable care in the preparation of this book, but makes no expressed or implied warranty of any kind and assumes no responsibility for any errors or omissions. No liability is assumed for incidental or consequential damages in connection with or arising out of information contained in this book. The Publisher shall not be liable for any special, consequential, or exemplary damages resulting, in whole or in part, from the readers' use of, or reliance upon, this material. Any parts of this book based on government reports are so indicated and copyright is claimed for those parts to the extent applicable to compilations of such works.

Independent verification should be sought for any data, advice or recommendations contained in this book. In addition, no responsibility is assumed by the publisher for any injury and/or damage to persons or property arising from any methods, products, instructions, ideas or otherwise contained in this publication.

This publication is designed to provide accurate and authoritative information with regard to the subject matter covered herein. It is sold with the clear understanding that the Publisher is not engaged in rendering legal or any other professional services. If legal or any other expert assistance is required, the services of a competent person should be sought. FROM A DECLARATION OF PARTICIPANTS JOINTLY ADOPTED BY A COMMITTEE OF THE AMERICAN BAR ASSOCIATION AND A COMMITTEE OF PUBLISHERS.

Additional color graphics may be available in the e-book version of this book.

Library of Congress Cataloging-in-Publication Data

Names: Eisenlohr, Pedro, editor.
Title: Forest conservation : methods, management and challenges / [edited by] Pedro Eisenlohr (State University of Mato Grosso (UNEMAT), Alta Floresta, MT, Brasil).
Description: Hauppauge, New York : Nova Science Publisher's, Inc., [2018] | Series: Environmental research advances | Includes bibliographical references and index.
Identifiers: LCCN 2018044672 (print) | LCCN 2018046378 (ebook) | ISBN 9781536145601 (ebook) | ISBN 9781536145595 (hardcover) | ISBN 9781536145601 (ebook)
Subjects: LCSH: Forest conservation.
Classification: LCC SD411 (ebook) | LCC SD411 .F59 2018 (print) | DDC 333.75/16--dc23
LC record available at https://lccn.loc.gov/2018044672

Published by Nova Science Publishers, Inc. † New York

CONTENTS

Preface		ix
Chapter 1	Rare Tree Species as Surrogates for Biodiversity in Conservation Decision-Making: What Are They and How to Select and Use Them? *Everton A. Maciel and Fernando R. Martins*	1
Chapter 2	Sustainable Management of Biodiversity in Woody Ecosystems: Biotechnology and Bioprospecting of Native Species from Monte Desert, Patagonia *Patricia Boeri, Lucrecia Piñuel, Daniela Dalzotto, Daniel Barrio, Maite Romero Alves, Marianelén Cedrés Gazo and Sandra Sharry*	41
Chapter 3	Forests and Brazilian Reptiles: Challenges for Conservation *André Felipe Barreto-Lima and Melina Soledad Simoncini*	67

Chapter 4	Prioritization of Areas for Permanent Preservation for Forest Recovery Aiming Landscape Connectivity *Emanuelle Brugnara,* *Vinícius de Freitas Silgueiro* *and Julio Cesar Wojciechowski*	**113**
Chapter 5	Conservation of Aleppo Pine Forest for Post Flood and Fire Plantings *Abdelaziz Ayari*	**153**
Chapter 6	Linking Agroforestry to REDD+ Activities in the Amazon: Implications for Biodiversity and Carbon Conservation *Pedro Manuel Villa, Sebastião Venâncio Martins,* *Silvio Nolasco de Oliveira Neto,* *Alice Cristina Rodrigues* *and Lucieta Guerreiro Martorano*	**171**
Chapter 7	Forest Conservation and Its Challenges in Tropical Africa *Tsegaye Tagesse Gatiso*	**207**
Chapter 8	Large Dams in the Amazon: Disconnecting the Social and Natural System *Elineide Eugênio Marques and Adriana Malvasio*	**225**
Chapter 9	Selection and Propagation of Native Tree Species for Improving Ecological Restoration *Sebastián Pablo Galarco, Maite Romero Alves,* *Patricia Boeri, Luciano Roussy, Marina Adema,* *Blanca Villarreal, María Valentina Briones,* *María de los Ángeles Basiglio Cordal,* *Tatiana Cinquetti, Diego Iván Ramilo* *and Sandra Elizabeth Sharry*	**239**

Chapter 10	Concepts and Methods in Environmental Suitability Modeling, an Important Tool for Forest Conservation	**269**
	João Carlos Pires-Oliveira, Leandro José-Silva, Diogo Souza Bezerra da Rocha and Pedro V. Eisenlohr	
About the Editor		**291**
Index		**293**

PREFACE

Forest Conservation: Methods, Management and Challenges offers to readers the opportunity to understand, consider and plan strategies that aim to conserve forest ecosystems around the world. This book presents ten chapters written by renowned researchers from Brazil, Argentina, Tunisia and Germany, offering to the scientific community – as well as to human society as a whole – important concepts, methods and gaps that we need to fill if we wish to preserve the Earth's forests.

The authors begin this collection by demonstrating how rare tree species could be a surrogate for biodiversity in conservation decision-making (Chapter One). Sustainable management of biodiversity in woody ecosystems is the theme of Chapter Two, followed by an interesting synthesis and discussion on challenges for conservation of forests and Brazilian reptiles (Chapter Three). Prioritization of areas for permanent preservation for forest recovery aiming at landscape connectivity (Chapter Four), conservation of Aleppo pine forests for post flood and fire plantings (Chapter Five), agroforestry and its connections to REDD+ activities in the Amazon (Chapter Six), forest conservation and its challenges in tropical Africa (Chapter Seven), large dams in the Amazon and their effects on the fauna (Chapter Eight) and selection and propagation of native tree species for improving ecological restoration (Chapter Nine) are themes deeply addressed in the next contributions, including interesting case studies. This book ends with an approach to environmental suitability modeling and its potential to support conservation decisions and ecological restoration programs in virtually any part of the world (Chapter Ten).

Forest Conservation: Methods, Management and Challenges is an important tool for students, researchers, decision-makers, governmental and non-governmental agencies that are interested in preserving different forest types in order to assure biodiversity conservation for current and future generations.

In: Forest Conservation
Editor: Pedro V. Eisenlohr

ISBN: 978-1-53614-559-5
© 2019 Nova Science Publishers, Inc.

Chapter 1

RARE TREE SPECIES AS SURROGATES FOR BIODIVERSITY IN CONSERVATION DECISION-MAKING: WHAT ARE THEY AND HOW TO SELECT AND USE THEM?

Everton A. Maciel[1] and Fernando R. Martins[2,]*
[1]Graduate Program in Plant Biology,
University of Campinas, Campinas, SP, Brazil
[2]Department of Plant Biology,
University of Campinas, Campinas, SP, Brazil

INTRODUCTION

Extinction rates have been increasing over the last 200 years, producing higher relative effects than the background rate (Dirzo et al. 2014; Ceballos et al. 2015). Reducing extinction rates partly depends on our ability to identify species that are most susceptible to anthropogenic and stochastic

---

[*] Corresponding Author Email: fmartins@unicamp.br.

factors; this is the starting point in biodiversity conservation (Margules and Pressey 2000). It should also be ensured that these species are properly represented in conservation planning (Mace 2004). Ideally, biodiversity conservation should assume that all species in a given region are represented in conservation planning, but nevertheless this is an unworkable task, especially in hyperdiverse tropical regions (Possingham et al. 2007). Alternatively, what has been done in practice is to use some species as a surrogate for biodiversity, such as endangered species, umbrella species and others (Lambeck 1997; Groves et al. 2002). However, little attention has been cast over rare species, although they represent a group with high vulnerability to processes triggering extinction (Tilman et al. 1994; Wamelink et al. 2014). Indeed, rare species are poorly represented in decision-making. They are the focus of this chapter. As protecting species from extinction is a key point for biodiversity conservation (Kukkala and Moilanen 2013), rare species should be seen as one of the focal species available to guide conservation decision-making (Synge 1980; Gaston 1994).

One way of addressing species rarity can be found in Rabinowitz (1981), who has taken into account population size, geographic range, and the habitat preference of species. Rabinowitz's rarity scheme has been used to identify rare species in plant communities (Sætersdal and Birks 1997; Izco 1998; Pitman et al. 1999; Broennimann et al. 2005; Caiafa and Martins 2010; Gauthier et al. 2010). While it has been criticized (Kunin and Gaston 1993), Rabinowitz's scheme has proven to be successful because it has easy-to-follow instructions and is flexible in such a way that it can be used for different applications in many cases (e.g., Gauthier et al. 2010; Knapp and Salomón 2010). The application of Rabinowitz's scheme generates a list with species classified into different rarity forms (see Caiafa and Martins 2010). Thus, the easy-to-follow instructions, the flexibility and resulting product of Rabinowitz's scheme provides a well-established combination that may help to reduce the gap in conservation-making (Maciel et al. 2016).

Nevertheless, rare species are seldom considered in conservation-making decisions, although it is admitted that their use could affect the Global Strategy for Plant Conservation (www.cbd.int/gspc/targets) because conservation actions may fail to protect rare species (Prendergast et al. 1993). We are concerned with this fact and assume that it has to do with the difficulty of producing a rigorous definition and selection of what rare species are (Izco 1998; Bevill et al. 1999). Although the scheme for attributing rarity forms to different species has been in existence for a long time (Rabinowitz 1981a), it has still been seldom used. Our focus in this chapter is to discuss the role of rare species in light of the biodiversity conservation theory.

However, we warn that some of the topics about rarity causes still open for discussion are beyond the scope of this chapter, although they are addressed in other studies, such as Gaston (1994, 2009). On the other hand, if you are seeking to discover how rare species can be used for biodiversity conservation, we encourage you to go forward into the full chapter. Here, our main goal is to describe two approaches for rare species classification: the first strictly complies with Rabinowitz's proposal (e.g., Caiafa and Martins 2010), and the second is also based on Rabinowitz, but has a few amendments (Gauthier et al. 2010). Additionally, we also discuss the need to insert rare species in decision-making based on the vulnerability principles (Margules and Pressey 2000). More specifically, in this chapter we address the following points:

1) Why have species become rare?
2) How should rare species be selected?
3) What are the consequences of using rare species for decision-making in biodiversity conservation?
4) What are the reasons for conserving rare species?

WHY HAVE SPECIES BECOME RARE?

A natural community is composed of species that are unequally successful, implying a whole range of abundances, in which some species predominate and others are rare (Whittaker 1965). If we plot the cumulative number of species of a community on the abscissa of a diagram and the cumulative number of individuals on the ordinate, we will observe something similar to a positive exponential curve (Gleason 1929). Although the exact shape of the curve depends on the different measures used to quantify abundance (for example relative abundance, number of individuals, biomass), the subsequent graphical analysis of this curve always divides the community into a small group of common species and a huge group of rare species (Henderson and Magurran 2010, Figure 1a).

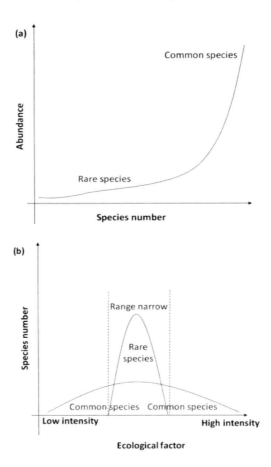

Figure 1. Schematic representation of rare and common species. (a) Tendency curve between the cumulative number of individuals (abundance) and the cumulative number of species (species number): in a community, many species have few individuals, and few species have many individuals. (b) Hump-shaped curves showing the distribution of a generalist (platykurtic curve) and a specialist (leptokurtic curve) species.

Now, if we plot the species abundance on the ordinate and some ecological resource/factor (for example pH or salinity) on the abscissa, we will observe a hump-shaped curve indicating that each species occupies a definite segment of the abscissa (Condit et al. 2013). This is because the withstood range of ecological conditions is particular for each species (Rabinowitz 1978; Pärtel et al. 2004; Pandit et al. 2009): a given species is not able to survive outside of the ecological conditions that it can withstand, that is, every species has its niche limits (Murray and Lepschi 2004). For some species a narrow range can mean restricted environmental specialization (Brown et al. 2003; Simon and Du Vall Hay 2003; Williams et al. 2009; Spitale 2012; Wamelink et al. 2014). Consequently, some species are common and show distribution over a wide range of ecological resources/factors, while many other species are rare and can only occur where they meet the restrict environmental conditions that they can cope with (Field and Coddington 1980; Wamelink et al. 2014, Figure 1b).

Many factors are considered causes of rarity (Field and Coddington 1980; Kunin and Gaston, 1993; Fiedler 1995; Izco 1998; Gaston 2009). For example, biotic interactions are important factors because some species are more successful in coexistence with other species in space-time (Sætersdal and Birks 1997; Mooney and Cleland 2001; Liancourt et al. 2005; Soberón 2007; Wolkovich et al. 2014). In addition, rarity can reflect the species ability to cope with abiotic stress (Rocap et al. 2003; Wamelink et al. 2014). Likewise, random factors, such as climatic oscillations, for instance, can restrict the species distribution (Dynesius and Jansson 2000). In short, biotic and abiotic drivers and habitat change can act as causes of species rarity.

Biotic Factors Shaping Distribution Patterns of Rare Species

Considering biotic factors as the drivers of species distribution is in accordance with the Eltonian niche theory, which states that a species' niche is limited by biotic interactions (Elton 1927). In a simplified form, from the viewpoint of this chapter, biotic interactions include both plant-plant and/or plant-other organism interactions. Biotic interactions can be positive, which tend to expand the species' niche, or negative, which tend to reduce the species' niche. Positive biotic interactions include facilitation, mutualism, commensalism, and protocooperation, whereas negative biotic interactions encompass interspecific competition (including allelopathy), herbivory, and parasitism (including pathogens). Since negative biotic interactions can potentially reduce the species' niche and contribute to its rarity, we give some explanation below.

Interspecific Competition

Rarity can probably arise when a species is at a great disadvantage compared to higher competitor species (Rabinowitz et al. 1984). When two species depend on one same resource, one of them has a particular characteristic that favors the supply that best meets its requirements, leaving behind the other species with low competition ability, which consequently suffers diminution in its population biomass (Price and Kirkpatrick 2009).

Parasitism

Pest pressures can reduce a species population in many different ways. For example, some natural enemies can cause a decline in population rates (Wills et al. 1997; Bachelot and Kobe 2013) and reduce abruptly the species range (Holt and Barfield 2009). In some case, natural enemy effects in rare species are more severe than in common species (Mangan et al. 2010). For example, rare species show slower growth in soil where pathogenic fungi are present, but this is not true for abundant species (Klironomos 2002). In addition, the removal of herbivorous species in a region reduced the mortality rate from 60% to 6.7%, and increased the reproductive success of *Cirsium pitcheri* (Torr. ex Eaton) Torr. & A.Gray (Asteraceae), a rare species in sand dunes around the western Great Lakes of Canada and the United States (Bevill et al. 1999).

Abiotic Factors Shaping Distribution Patterns of Rare Species

Climate, especially precipitation, temperature, and seasonality, is amongst the most important abiotic factors controlling species' geographic distribution, but soil pH, salinity, and phosphorus content have also an important role (Dajoz 1971; Gaston and Fuller 2009; Condit et al. 2013). Considering abiotic factors/resources as capable of limiting species' distribution is consistent with Grinnellian niche theory, according to which the observed species distribution is limited in space by the variation of these factors/resources (Grinnell 1917). For example, some plant families in China show different spatial distribution patterns of rare and common species in response to climatic variation (Liu et al. 2017). In the Brazilian Atlantic rainforest, many tree taxa do not cross south of the Capricorn tropic line, thus defining species with tropical and subtropical niches (Giehl and Jarenkow 2012). In short, climate can differently limit geographic ranges of rare and common species (Bachelot et al. 2015). Many other abiotic factors/resources can limit a species' geographic range, such as soil characteristic. For example, in an extensive research based on 8,000 plots in the Netherlands, Wamelink et al. (2014) investigated 973 species (of which 190 were considered rare) and concluded that rare species had the smallest geographic range in responding to 19 of 20 abiotic factors (e.g., soil phosphorus, pH, and others). Also, the rarer the species the less is the withstood range of these abiotic factors/resources (Wamelink et al. 2014). This is because rarity can take different forms, and each of them has different withstood limits (Rabinowitz 1981b).

Rarity and the Present State of Knowledge

There is presently a great effort to understand the causes of natural rarity patterns (Field and Coddington 1980; Kunin and Gaston 1993; Fiedler 1995; Murray and Lepschi 2004; Gaston 2009). Many biotic and abiotic factors/resources are known to be very important in delimiting species geographic range, but many other variables are thought to participate in the shaping of species present distribution, such as, for instance, seed dispersal limitation, demographic and environmental stochasticity, random catastrophes, climatic oscillations, biogeographic processes etc.

Above we gave just a few examples, picked up among many others in the literature (Gaston 2009). Anyway, we just want to make it clear that natural rarity can occur due to a number of reasons. Consequently, according to Rabinowitz (1981), species have different ways of being rare. Caiafa and Martins (2010) have attributed characteristic names to these different forms of rarity. Some species in communities are stenoecious (occur in a specific habitat), whereas some are euryoecious (occur in various habitats); some are stenotopic (have a limited geographic distribution), while others are eurytopic (have a wide geographic distribution). Species with a small local population are commonly characterized as singletons (one individual in a sample) or doubletons (two individuals in a sample) and were called scarce, whereas species with a large local population were called abundant by Caiafa and Martins (2010). In the following section, we will describe the procedures to classify species' rarity forms.

HOW SHOULD RARE SPECIES BE SELECTED?

Many investigations have consistently shown that natural communities have a rarity pattern (Field and Coddington 1980; Fiedler 1995), generally presenting many scarce species and few abundant ones (Pires and Prance 1977). One now well-accepted scheme dealing with rarity is Rabinowitz's (Rabinowitz 1981; Rabinowitz et al. 1986), who has proposed seven forms of rarity. Although Rabinowitz's scheme does not aim to explain the causes of rarity (Izco 1998; Bevill et al. 1999), it provides an interesting theoretical framework for classifying rare species. Rabinowitz (1981) and Rabinowitz et al. (1986) have originally elaborated their scheme based on the British flora, considering that species can assume a continuous gradation in geographic range, population size, and habitat specificity. Each one of these three parameters can assume two states: (a) wide or narrow geographic range, (b) abundant or scarce local populations, and (c) many or a single habitat. In consequence, eight classes are generated, one of which has species with wide geographic range and abundant local populations in many different habitats – these are the common species. The other seven classes are the forms of rarity. For example, the rarity form 7 is attributed to a species whose geographic range encompasses a narrow latitudinal range (e.g., Caiafa and Martins 2010), in which its populations are sampled with only one or two individuals (e.g., Pitman et al. 1999) and occur in a specific habitat (Figure 2).

Following Rabinowitz's scheme, Gauthier (2010) proposed an alternative approach (hereinafter referred to as Gauthier's scheme), in which there are a number of important differences. In Gauthier's scheme, Rabinowitz rarity parameters of geographic range, local abundance and habitat specificity are translated into regional responsibility, local rarity and habitat vulnerability, respectively. Regional responsibility (Figure 2) takes political delimitation into account. In Gauthier's scheme, a political delimitation is a first-rank division of a country into political units, such as provinces in France, departments in Colombia, or states in Brazil. For example, Gauthier et al. (2010) has used the number of French Provinces with the species presence to quantify the regional responsibility for a species: the smaller the number of provinces with the species presence the higher the regional responsibility. Local rarity is expressed as the number of localities with the species presence within an area of interest: the smaller the occurrence number of the species the greater the species local rarity. Habitat vulnerability is inferred from the potential human threatens and pressure upon an area that is potentially suited to human use, such as lands appropriated to husbandry or mining or any productive vocation: the greater the potential human use of an area the greater the habitat vulnerability. Considering each one of these parameters, a score is attributed to each species, and then the scores of each species are rearranged, so that the results of the combination of regional responsibility, local rarity and habitat vulnerability are used to compose the priority species list. Each species list attributes a score ranging from 1 for low priority species to 5 for high priority species (Figure 2).

Although each of these methods has some particularities, both approaches still conserve the capacity of multiscale selection, because the three parameters operate on different spatial scales, such as local and regional level (Hartley and Kunin 2003). Furthermore, both approaches can provide a list of species graded according to the seven forms of rarity (e.g., Caiafa and Martins 2010), or a species list with different proprietary scores (Gauthier et al. 2010), and ultimately show which species are rarer in the community.

Indeed, a number of studies have shown that rare species can be selected by both Rabinowitz's and Gauthier's scheme. Some examples of these studies include the flora of the Canadian Boreal Plain in the province of Saskatchewan (Kricsfalusy and Trevisan 2014, Figure 2a); the Central Spanish wet-grassland habitats of the Sierra de Guadarrama (Rey Benayas et al. 1999, Figure 2b); the Norwegian flora (Sætersdal and Birks 1997, Figure 2e); the Languedoc-Roussillon region and Causse Méjean in the Mediterranean part of France (Gauthier et al. 2010, Figure 2d); the flora of Switzerland (Broennimann et al. 2005, Figure 2e), the Amazonian flora in Manu, Peru (Pitman et al. 1999, Figure 2f); the Brazilian Cerrado Province-Amazon transition zone (Maciel et al. 2016, Figure 2g); and the Brazilian Atlantic rainforest (Caiafa and Martins, 2010; Figure 2h). In Figure 2, all studies show different forms of rarity according to both Rabinowitz's and Gauthiers' schemes.

However, the different procedures adopted by the different authors to fulfil the criteria of each one of these schemes make it impossible to draw direct, exact comparisons between them. Yet, it becomes clear that rarity can express different grades or forms in nature. In addition, almost every paper cited here includes a section dedicated to highlight the importance of rare species to biodiversity conservation. In the following paragraphs, we describe how each of the two rarity schemes can be dealt with in practice, taking tree species of the Brazilian savanna as an example.

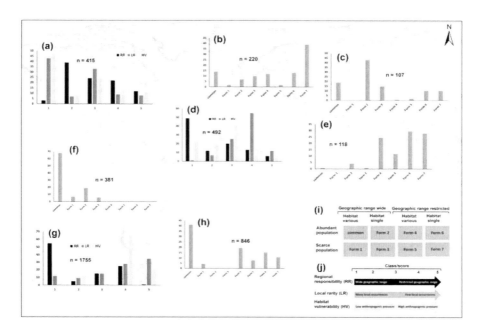

Figure 2. Surveys that have applied Rabinowitz's (black-bar histograms: b, c, e, f, h) or Gauthier's (a, d, g) scheme in different regions of the world. The number of species (n) in each survey is shown. Regional responsibility (RR), local rarity (LR) and habitat vulnerability (HV) are Gauthier's parameters. Rarity increases in the abscissa from left to the right, and the ordinate axis represents the percentage of species in each rarity category. Both Rabinowitz's (i) and Gauthier's (j) schemes are shown.

The Brazilian savanna receives the local name of cerrado. The Cerrado Phytogeographic Province occupies the Brazilian Central Plateau, and stretches towards the northwest and southeast (Figure 3c, 3d). However, the cerrado vegetation has many disjunctions immersed in the Amazon, Caatinga, and Atlantic Forest provinces. The cerrado tree flora contains many rare species scattered not only in the Cerrado Phytogeographic Province, but also in neighbor formations (Castro et al. 1999; Françoso et al. 2016). In addition, the Cerrado Phytogeographic Province has high endemism (Mittermeier et al. 1998; Simon and Proença 2000; Forzza et al. 2012; Zappi et al. 2015), which is very important for rare species classification schemes (Gauthier et al. 2010). Furthermore, the Cerrado Phytogeographic Province has very diverse habitats, which are expressed as both a strong floristic provincialism (Ratter et al. 1996) and a great variation of the physiognomy and wood biomass from typical grasslands to forest vegetation driven by soil properties (Arens 1958; Goodland and Pollard 1973; Lopes and Cox 1977), fire (Coutinho 2006) and rainfall (Miranda et al. 2014). To give an example of how to use Rabinowitz's and Gaulthier's schemes we considered tree species found in the NeotropTree database (Oliveira-Filho 2017), which is an open database that also furnishes information on many factors associated with the species, such as, for example, occurrence, habitat, province/state, and many others (Eisenlohr and Oliveira-Filho 2016).

CLASSIFYING SPECIES INTO THE SEVEN FORMS OF RARITY OF RABINOWITZ'S SCHEME

Geographic Range

The geographic range of a species considers the geographic distribution of all populations of this species in order to set the extreme points within which they occur. Traditionally, geographic range has been addressed by measuring the percentage of plots with a species presence in relation to the total of plots set across a region (e.g., Sætersdal and Birks 1997). This is the same as constancy (Braun-Blanquet 1951, 1979) and was used by Scudeller et al. (2001) to assess the geographic range of tree species in the southern sector of the Brazilian Atlantic rainforest. Species occurring in less than 10% of all plots are considered to have a restricted geographic range (Sætersdal and Birks 1997), being rare due to stenotopy. A second way to assess geographic range is the latitudinal amplitude, that is, the "latitudinal difference in minutes between the two most distance (north-south) surveys in which species I occurs" (Scudeller et al. 2001).

Rare Tree Species as Surrogates for Biodiversity ...

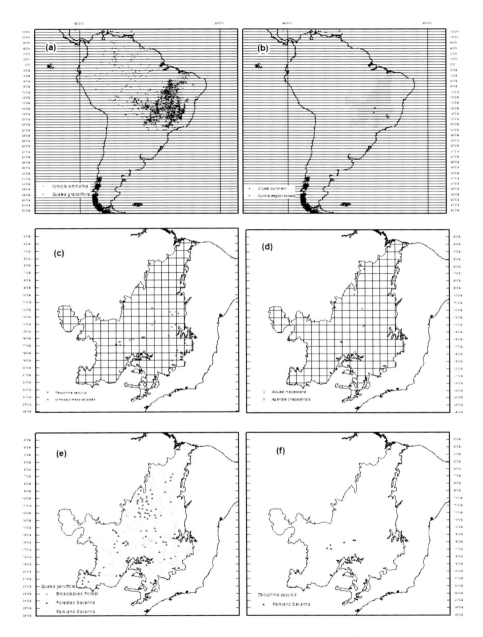

Figure 3. Rabinowitz's forms of rarity of tree species in the Brazilian Cerrado Province: species with wide (a) and restricted (b) geographic range, abundant (c) and scarce (d) populations, and habitat-indifferent (e) and -specialist (f) species.

Still another way to assess geographic range, which we adopt here, is to consider the number of latitudinal quotas where the species has been observed to occur (e.g., Caiafa and Martins 2010). To do that, all occurrence points of a species must be taken into account by performing a detailed search in regional and national databases and data collection, such as the Global Biodiversity Information Facility (www.gbif.org). Subsequently, geographical information system techniques (GIS), such as ArcMap (ESRI 2011) or QGIS (https://www.qgis.org/en/site/), should be applied to check each geographic coordinate and exclude nonsense coordinates.

In our example, we consider the entire continent of South America as a potential occurrence area of tree cerrado species, since some of them occur throughout the continent (Figure 3a and b). Thus, we consider that species occurring on only one or two latitudinal quotas (e.g., Caiafa and Martins 2010) of all the possible ones across the whole South America have a restricted geographic range (stenotopic species). In our example, *Qualea grandiflora* Mart. (Vochysiaceae) and *Xylopia aromatica* (Lam.) Mart. (Annonaceae) have wide geographic ranges, occurring on more than two latitudinal quotas, that is, they are eurytopic species (Figure 3a). Looking more carefully, you notice that *X. aromatica* is widely spread from the latitude parallel of 8o north to the parallel of 24o south (Figure 3a). In contrast, *Clusia burchellii* Engl. (Clusiaceae) and *Myrcia aegiphiloides* Mattos (Myrtaceae) have a restricted geographic range, not exceeding two latitudinal quotas, and as a result they would be considered rare species due to stenotopy (Figure 3b).

Population Size

Numerical abundance (density, the number of individuals per unity area) is a metric used to characterize the population size of a species (Guo 2003). In ecological studies, singletons (species sampled with only one individual) and doubletons (species sampled with two individuals) are commonly considered rare due to scarcity (Pitman et al. 1999; Caiafa and Martins 2010). Unlike the geographic range, which considers the totality of possible geographic quotas on which a species could be observed, the quantification of a species population size considers a surface area of interest, which, in our case, is the Brazilian Cerrado Phytogeographic Province. So, the fundamental information permitting to classify a species as abundant or scarce comes from phytosociological surveys. For example, considering various phytosociological samples Caiafa and Martins (2010) attributed forms of abundance to Brazilian Atlantic forest tree species. However, as we do not always know species local abundance, grouping a species occurrence in grid cells can be an effective alternative. In this chapter we adopt this alternative by establishing a grid with about 160 cells of 1° latitude S by 1° longitude W covering the Cerrado Province (e.g., Simon and Proença 2000; Ratter et al. 2003). Then, we calculated the number of occurrences of a species per grid cells to attribute the species abundance: we consider as rare those species with only one or two occurrences by grid cells. For example, in our grid *Tibouchina papyrus* (Pohl) Toledo (Melastomataceae) and *Mimosa pithecellobium* Benth. (Fabaceae) (Figure 3c) exceed two occurrences by grid cells, therefore are not considered rare species. On the other hand, *Aiouea macedoana* Vattimo-Gil (Lauraceae) and *Agarista chapadensis* (Kin.-Gouv.) Judd (Ericaceae) do not have more than two occurrences in any grid cell, and are consequently rare species due to scarcity (Figure 3d). However, this approach has some bias, because many studies did not consider some requirements – i.e., sample size variation, size of the smallest individual to be sampled, and sampling method – a flaw that can lead to biased interpretations (Caiafa and Martins 2007).

Habitat Specificity

Some species occupy an extreme position as generalists occurring in a wide range of different habitats, whereas others occupy the other extreme as specialists supporting only a strict set of conditions (Pitman et al. 1999; Pandit et al. 2009). Being a specialist is considered a form of rarity (Rabinowitz 1981), that is, rarity due to stenoecy (Caiafa and Martins 2007). As a consequence of the high heterogeneity of the Cerrado Phytogeographic

Province environment on different scales (Goodland and Pollard 1973; Ribeiro and Walter 2008), great differences in physiognomy and species composition impose a strong provincialism (Ratter et al. 1996). Here, we take for grant that the phytophisionomy is a reliable indicator of environmental conditions (Coutinho 1978). Therefore, we consider that species occurring exclusively in one phytophysiognomy are habitat specialists, being rare due to stenoecy. For instance, *Qualea* is a genus that occurs in every Cerrado habitat type, and *Q. parviflora* Mart. (Vochysiaceae) occurs in more than one phytophysiognomy, being a euryoecious species (Figure 3e). Differently, *T. papyrus* is considered a habitat specialist because it only occurs in Parkland Savanna, thus being a stenoecious species (Figure 3f).

Rabinowitz's Forms of Rarity

Table 1. Rabinowitz's forms of rarity

Species	Geographic range		Population size		Habitat		Rarity form
	Wide	Restricted	Abundant	Scarce	Various	Single	
Sp 1	x		x		x		common
Sp 2	x			x	x		Form 1
Sp 3	x		x			x	Form 2
Sp 4	x			x		x	Form 3
Sp 5		x	x		x		Form 4
Sp 6		x		x	x		Form 5
Sp 7		x	x			x	Form 6
Sp 8		x		x		x	Form 7
(...)							
Sp n		x	x		x		Form f

If we know the geographic range, the local abundance of populations, and the habitat specificity of a species, we can attribute its form of rarity. By doing this for each species, we can produce a list of species according to their forms of rarity. This is done by constructing a table whose first column has the species names, the next three columns are split into the two possible states of each of the three rarity parameters, and the last column is the resulting form of commonness or rarity (Table 1).

CLASSIFYING SPECIES ACCORDING TO GAUTHIER'S SCHEME OF REGIONAL RARITY

Regional Responsability (RR)

This is a geographic criterion used to measure a species' geographic range (Schmeller et al. 2008). To attribute the regional responsibility for a species, the species occurrence inside and outside the focal area is compared. Gauthier et al. (2010) have proposed considering the number of the states/provinces both within and outside the focal area as a way to quantify the species occurrence. Three possible results can be expected: (i) the species does not occur outside any province; (ii) it occurs in a single outside province; or (iii) it occurs in many outside provinces (Table 2). Therefore, a species occurring only in the focal area or only in a province outside the focal area is considered of high regional responsibility (RR), but a species occurring in many states/provinces is considered of low RR (see Table 2). In our example, the Brazilian Cerrado Phytogeographic Province encompasses ten political divisions or states (see Figure 4a and b), but there are many other Brazilian states outside the Cerrado Province. The occurrence of some species, such as, for example, *X. aromatica* and *Q. grandiflora*, has been recorded, respectively, in six and four external states (Figure 4a). For this reason, these species are considered to be of low RR, thus receiving scores of 2 and 3, respectively. In contrast, *Aiouea piauhyensis* (Meisn.) Mez (Lauraceae) (Figure 4b), which has been recorded in only one external state, receives a score of 4. Species that, such as *C. burchellii* (Figure 4b), do not occur outside the states included in the Cerrado Province are given the score 5, which is the highest priority score.

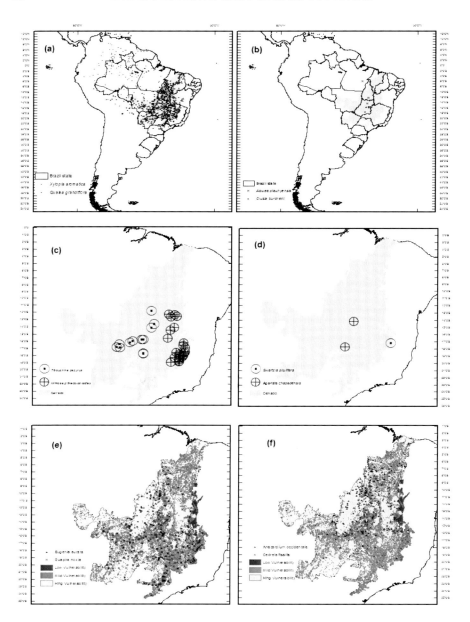

Figure 4. Gauthier's rarity categories of tree species in the Brazilian Cerrado Province: species with low (a) and high (b) regional responsibility, low (c) and (d) high local rarity, and low (e) and high (f) habitat vulnerability.

Table 2. Gauthier's scheme and proprietary score taking the Cerrado Province as the focal area

Criterion	Score/class	Criterion application
Regional responsibility (RR)	5	Species that occur only in the Cerrado Province
	4	Species that occur in 1 or 2 states outside the Cerrado Province
	3	Species that occur in 3 or 4 states outside the Cerrado Province
	2	Species that occur in 5 or 6 states outside the Cerrado Province
	1	Species that occur in 7 or more states outside the Cerrado Province
Local rarity (LR)	5	Species that occur in only one Cerrado site
	4	Species that occur in 2 to 5 Cerrado sites
	3	Species that occur in 6 to 10 Cerrado sites
	2	Species that occur in 11 to 30 Cerrado sites
	1	Species that occur in more than 30 Cerrado sites
Habitat vulnerability (HV)	5	Species that occur *exclusively* in areas with *HIGH* agricultural aptitude. *HIGH* is the result of a combination of two factors (1) land topography flat (0-3% declivity) or undulated (3-6% declivity) + (2) precipitation > 900 mm.
	4	Species that occur *simultaneously* in areas with *HIGH* + *MID* agricultural aptitude. *MID* is result of only one factor (1) land topography flat (0-3% declivity) or undulated (3-6% declivity) *or* (2) precipitation > 900 mm.
	3	Species that occur *exclusively* in areas with *MID* agricultural aptitude.
	2	Species that occur *simultaneously* in areas with *HIGH* + *LOW* agricultural aptitude or *MID* + *LOW* or *HIGH* + *MID* + *LOW*. *LOW* is result of land topography flat (>6% declivity) + precipitation < 900 mm.
	Score/class	Criterion application
	1	Species which occur *exclusively* in areas with *LOW* agricultural aptitude.

Local Rarity (LR)

Gauthier's scheme proposes to consider the known number of occurrences in a focal area to quantify LR, since the greater the number of occurrences the more abundant the species populations are. Luckily, the NeotropTree databank provides the number of sites with presence of each species in a region. To attribute a score of local rarity to a species, we computed each site surveyed in the Brazilian Cerrado with presence of a species "s" as one presence of species "s." Therefore, the smaller the number of presences the higher the LR score of species "s." In our example, *T. papyrus* and *Mimosa pithecolobioides* Benth. (Fabaceae) (Figure 4c) occur in more than five sites in the Cerrado Province (Table 2). Consequently, these species are rated with the low score of 1 for LR. However, *Swartzia pilulifera* Benth. (Fabaceae) and *Agarista chapadense* (Kin.-Gouv.) Judd (Ericaceae) have been observed in just one and two sites, respectively, and both are given the LR high scores of 4 and 5, respectively.

Habitat Vulnerability (HV)

The vulnerability of a species can be linked to both stochastic processes and land use (Gauthier et al. 2010). In the case of land use, good conditions of soil, climate, and topography should be considered as threatens because they may attract land use by agriculture, wood exploitation, or cattle handling, for example (Wilson et al. 2005). These practices can act synergistically with other factors, ultimately causing species extinction (Laurance and Useche 2009). In the Brazilian Cerrado, for example, the appropriateness of large extensions for husbandry is well known (Pereira and Lombardi-Neto 2004; Ladeira-Neto 2013). Large extensions of the Cerrado Province have flat to slightly hilly topography, and annual rainfall is equal to or greater than 900mm. As agriculture is one of the human activities imposing high pressure on the environment (Monteiro et al. 2017), agricultural aptitude as an important factor in species vulnerability is perfectly defensible. Here, in our example, we use data from the Brazilian

Geological Service (Brasil 2017) for three soil declivity categories, and Global Climate Data - WorldClim (www.worldclim.org) for three precipitation categories (Table 2, Figure 4e and f). We combined the data and generate HV layers for different categories of vulnerability (Table 2, Figure 4e and f). Then, we attributed the habitat vulnerability to each species. Species occurring in all vulnerable categories are considered to have low HV; species restricted to habitats with high or median vulnerability categories are considered to have high HV. In our example, *Eugenia aurata* O. Berg (Myrtaceae) and *Guapira noxia* (Netto) Lundell (Nyctaginaceae) occur in all vulnerability categories and thus are given a low score for HV (Figure 4e). However, *Cedrela fissilis* Vell. (Meliaceae) occurs only in high categories of vulnerability and is given a high score for HV (Figure 4f). As an example in between, we took *Anacardium occidentale* L. (Anacardiaceae), which occurs in high and median categories of vulnerability and is given a higher score for HV than those of *E. aurata* and *G. noxia,* but lower than *C. fissilis.*

Rating Species Conservation Priority According to Gauthier

With information about regional responsibility, local rarity, and habitat vulnerability for each species, the next step is to attribute a priority rank to each species, in order to produce a list of species rated according to their scores (Table 3). This is achieved by arranging the species in a table with six columns: the species in the first column; each one of the three parameters in the following three columns; the computation resulting from these three parameters (final score) in the fifth column; and finally the sixth column (priority rank) is the ranking of the species according to conservation priority. Gauthier has proposed two ways of computing the final score of each species: (1) as a simple average of the three parameters ([RR+LR+HV]/3); or (2) as a weighted average resulting from the assigning of a weight to regional responsibility, with the intent to increase the priority rank of the species that only occur in the focal area. In our example, we assign a weight of 2 to RR; then, the species final score was calculated as

(2RR+LR+HV)/4. Finally, the combination of the species' score for each parameter is used to produce an overall ranking in order to make a hierarchical arrangement of the species according to their conservation priority (Table 3).

Table 3. How the scores can be combined to rank species according to Gauthier's criteria (RR = Regional responsibility, LR = Local rarity and HV = Habitat vulnerability)

Species	RR	LR	HV	Final score	Priority rank
Sp 1	5	4	5	4.75	1
Sp 2	4	5	5	4.5	2
Sp 3	5	4	4	4.5	2
Sp 4	5	2	5	4.25	3
Sp 5	5	2	4	4	4
Sp n	3	4	5	3.75	5
(...)	5	1	4	3.75	5
Sp n	2	5	5	3.5	6

Forms of Rarity vs. Regional Priority: The Cerrado as an Example

Having contrasted what is meant in Rabinowitz's and Gauthier's schemes, we now propose the following questions:

1) Are Rabinowitz's and Gauthier's schemes coincident in attributing similar categories of rarity to the same species?

2) Are both schemes coincident in indicating the same geographic area with similar number of rare species?

3) Is there congruence between the composition of tree species selected for both methods?

Here, we considered that our areas having high richness of rare species are hotspots of rarity because hotspots can be considered areas whith exceptional concentrations of endemic species and habitat loss (Myers et al. 2000), characteristics found in both Rabinowitz's and Gauthier's rare species. To answer these questions, we built up a new dataset by picking up the 915 cerrado tree species and their respective coordinates from the NeotropTree database (Oliveira-Filho 2017) and classified the rarity of these species according to both Rabinowitz's and Gauthier's schemes as shown above.

Then, we delimited the Cerrado Province by adopting the concept given by the Environment Ministry of Brazil (http://www.mma.gov.br/biomas/cerrado), since it establishes the "official" Cerrado Province limits considered for decision-making in Brazil. We dividided the Cerrado Province into grid cells of 1° latitude x 1° longitude and computed the species number within each cell to obtain the total species richness per cell. Species richness is an important measure that underlies ecological studies and conservation decisions (Gotelli and Colwell 2001) and has been proposed as one of the key variables of biodiversity in order to monitor biodiversity conservation (Pereira et al. 2013).

We also computed for each cell the rare species according to both Rabinowitz and Gauthier, so that we had the total species richness, the number of rare species according to Rabinowitz (Rabinowitz's species richness) and the number of rare species according to Gauthier (Gauthier's species richness). Rabinowitz's species richness included species with any rarity form, but not common species. In the list of Gauthier's rare species we included those with a final score ≥3. Using higher scores to select species has the advantage of selecting high priority species among all rare species. Knowing these three variables in each cell enabled us to investigate whether a species classified as rare according to Rabinowitz's scheme is also classified as rare in Gauthier's scheme. Also, it allowed us to assess whether both schemes indicate a similar number of rare species in a same geographic area. Finally, it renders it possible to test the congruence between total species richness and Rabinowitz's + Gauthier's rare species richness (e.g., Orme et al. 2005).

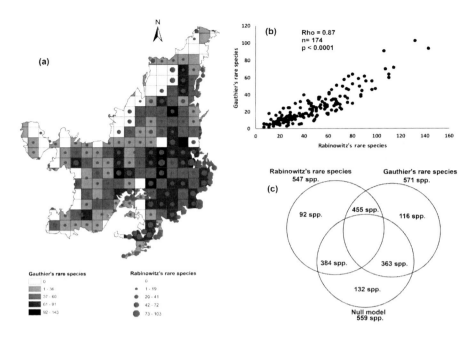

Figure 5. Congruence in rare species selected by Rabinowitz's and Gauthier's scheme in the Brazilian Cerrado Province. (a) The number of Rabinowitz rare species is congruent with the number of Gauthier rare species in each grid cell of 1° latitude south x 1° longitude west. (b) Scatterplot showing a strong direct correlation (Spearman coefficient) of the number of Gauthier rare species with the number of Rabinowitz rare species in each one of the 174 grid cells. (c) Venn diagram showing a greater congruence between Rabinowitz's and Gauthier's scheme than between each one of them and species drawn at random from the list of 915 total tree species.

Our results show that there is a strong spatial congruence between Rabinowitz's and Gauthier's rare species (Spearman coefficient rho = 0.87; $p < 0.0001$) (Figure 5a and b). To be sure that the observed congruence between the two schemes was not due to randomness, we randomly drew 559 species among the total 915 Cerrado species (null model) using the *sample function* in R (R Core 2017). Then, we used the Venn diagram (e.g., Zhao et al. 2016) in order to assess if the congruence of species selected by each scheme differs between each scheme and between them and the null model. Our Venn diagram shows that the congruence between both schemes is greater than the congruence bewteen both schemes and the null model (Figure 5c). To test this assumption we performed a chi-square test to check whether the proportion of species is congruent between the two schemes and whether both schemes are different from the null model. Our results confirm our assumptions: there was a significant difference between the two conditions ($X^2_{0.05;2}$ = 11.602, p < 0.003). It is therefore likely that Rabinowitz's scheme and Gauthier's scheme can indistinctly be used to assess species rarity.

A possible explanation for this might be that Gauthier's scheme is based on Rabinowitz's, thus yielding similar results. However, the interpretation of the results is quite different between the schemes (Figure 5c). Gauthier's scheme directly inserts a political, administrative approach in addition to the anthropic factor represented by the land aptitude. Therefore, we believe in combining both schemes for selecting rare species to have more effective results for use in decision-making.

WHAT ARE THE CONSEQUENCES OF USING RARE SPECIES FOR DECISION-MAKING IN BIODIVERSITY CONSERVATION?

In this section, our primary focus is to show how rare species may be used in biodiversity conservation strategies. For this we used rare species simultaneously selected by Rabinowitz's and Gauthier's scheme. In addition, we also discuss the fragility embedded in conservation approaches that do not consider using rare species. We argue that considering rare species for decision-making in biodiversity conservation, besides protecting directly rare, vulnerable species, plays an important role in protecting all other species.

Rare Species as a Substitute for Biodiversity Conservation: An Example in the Cerrado

We think that considering a species group as a surrogate for the entire biodiversity is efficient only if this group is directly related to the other groups that should be protected by conservation actions. Evaluating such an efficiency may be achieved by using simpler methods, such as the correlation between surrogate hotspots and hotspots of all species (Vera et al. 2011). Thus, the higher the degree of correlation between the two kinds of hotspots the more likely the existence of an efficient surrogate. In this regard, rare species are a very interesting group that has shown great affinity with areas of high species richness (Lennon et al. 2011; Spitale 2012; Villalobos et al. 2013).

Based upon this premise, we questioned whether the number of cerrado rare species is related to the number of all other species. If it is, we expected to find a direct correlation between the number of rare species and the number of all species in a site, since an efficient surrogate is considered to have a high correlation with what it represents (Xu et al. 2008). To attain this task, we considered both the rare and the total number of species in each cell of $1°$ latitude south x $1°$ longitude west in our grid of the Cerrado Province. To select the rare species in each cell we took the species classified as rare by both Rabinowitz's and Gauthier's schemes (Figure 6). Then, we performed a correlation analysis with Spearman coefficient between the total number of species and the number of rare species in each grid cell (see Pearman and Weber 2007). We obtained a strong correlation (rho = 0.84; p = 0.0001), indicating that conserving rare species can conserve all other species too (Figure 6). The congruence between rare species and total species number is evident in both individual cells and the dispersion diagram. Our results confirm that rare species occurrence can be associated with the total number of species (Lennon et al. 2011; Villalobos et al. 2013). Therefore, simply by conserving rare species it is possible to conserve all species.

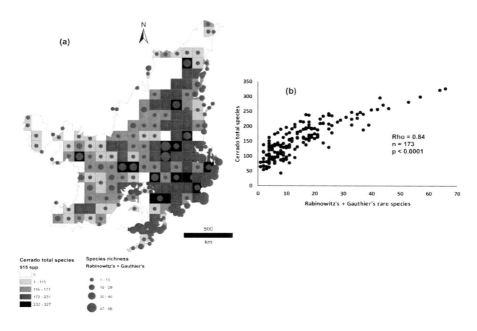

Figure 6. Congruence between rare tree species selected by both Rabinowitz's and Gauthier's scheme and the total tree species in the Brazilian Cerrado Province. (a) The number of rare species is strongly congruent with the total number of species in each grid cell of 1° latitude south x 1° longitude west. (b) The number of rare species is strongly directly correlated (Spearman coefficient) with the total number of species in each one of the 174 grid cells.

Rare Species in Decision-Making: The Achilles' Heel

The main criticisms related to using rare species for decision-making are focused on two points. One major drawback of this approach is that using rare species in decision-making does not only impact species in strategic international red lists such as IUCN's, but also others such as wide distributed species (McIntyre 1992). Despite this, consider a conservation planning that takes into account only species with restricted distribution to specific habitats. These planning would have great chances of not capturing the distribution area of the species that are widely distributed. Many threatened species of IUCN red list and also common species qualify as widely distributed; then, there would be many chances that these wide distributed species will not adequately be addressed. Another problem with this approach is that it fails to take the incomplete knowledge of distributions of species into account (see Field and Coddington 1980; Kunin and Gaston 1993). In fact, there is still a great void of data on biodiversity (Feeley and Silman 2011; Feeley 2015), and, as a consequence, the rarity of a species could reflect this knowledge gap. Here are some arguments in order to show you why these limitations do not invalidate the use of rare species in conservation planning.

In favor of rare species, we say that using them in conservation decision-making is not an attempt to neglect IUCN species or other species groups that can be used in decision-making, but quite the opposite, since the planning for effective biodiversity conservation can and should be maximized with a stratified surrogate that combines rare, threatened and common species (Maciel and Eisenlohr 2016). The knowledge gap concerning taxon distribution is commonly known as Linnean shortfall (Brito 2010; Hubbell 2013). This is a known problem, since rare species may reveal to occur more often in regions that are still poorly or not yet surveyed (Lõhmus 2015). Hence, the present lists of rare species might soon become obsolete (Field and Coddington 1980). However, conservation should be responsible for protecting species in all places (Arponen 2012). Therefore, the discovery of other rare species in the future should not invalidate strategies that are implemented in the present. Moreover, the high extinction rates that species in general are now experiencing need to be urgently resolved, even in the face of incomplete knowledge (Ceballos et al. 2015). This is especially urgent in a region of high richness, such as the tropics, particularly the Neotropics, which still demand hard, intense field work in order to allow decision-makers to have a reasonable knowledge about biodiversity (Cofre and Marquet 1999; de Lima et al. 2015). However, some authors, such as Corlett (2016), hope that missing data are due to natural rarity and not to a real lack of knowledge about the living taxa.

We should bear in mind that no system for classifying species is failure-free. In fact, even the IUCN red list, as one of the world's most recognized and trusted classifying systems, may be affected by data quality (Eaton et al. 2005). In addition, there can be no assurance that well-known species groups will render more successful conservation strategies than the little-known groups (Lentini and Wintle 2015). So, we must think about rare species as groups that are highly vulnerable to anthropic and stochastic process and seek hardly to conserve both the rare species and their habitats, because a proactive conservation policy is less costly than habitat restoration (Haslouer et al. 2005).

It is important to render it clear that conservation approaches have chances of something going wrong, thus affecting the achievement of their objectives (Game et al. 2013). However, our example with the cerrado tree flora shows that the amount of rare tree species is strongly directly related to all tree species number, leading to the conclusion that conservation planning designed for rare species can further reinforce the conservation of all species in a region. Hence, even rare species that are not included in international rare species list, such as IUCN, should be used in conservation decision-making (Sólymos and Feher 2005; Kricsfalusy and Trevisan 2014; Maciel et al. 2016; Le Berre et al. 2018).

WHAT ARE THE REASONS FOR CONSERVING RARE SPECIES?

Species richness is one of the factors that must be taken into account in decision-making (Lewandowski et al. 2010), because species loss results in biodiversity loss (Soulé et al. 2003). Thus, to reduce such biodiversity loss we need to reduce loss of species populations, since reduced population is the main cause of extinction (Linklater 2003). Put into other words, rarity is the precursor of extinction, and as decision-making in conservation should reflect the requirements of each species (Mace 2004; Kricsfalusy and Trevisan 2014), greater attention should be paid to rare species. On the one hand, rare species lack capacity to colonize environments that have already been altered by human activities (Pilgrim et al. 2004). On the other hand, besides being more vulnerable to continued habitat loss, rare species are a reliable surrogate for the local species pool (Lõhmus 2015), and also good indicators of specific (Pitman et al. 1999; Pärtel et al. 2004; Hubbell 2013) and/or rare (Pärtel et al. 2004; Pilgrim et al. 2004) habitats.

In accordance with the vulnerability principle, a greater risk of a species to be lost due to anthropic and stochastic processes means a greater priority for conservation (Margules and Pressey 2000; Wilson et al. 2005; Kukkala and Moilanen 2013). In consequence, protecting the most vulnerable species is the main point in conservation planning (Margules et al. 2002). Our results show that the cerrado has many rare species and they occur exclusively in areas with high agricultural aptitude, that is, areas combining flat and slightly hilly topographies with high precipitation. In addition of being considered an important factor in species loss, agriculture growth is also expected to reach strong expansion. Thus, based on the vulnerability principle (Margules and Pressey 2000), rare species need to be urgently conserved.

Ultimately, there is an urgent need to debate purposes related to conservation of rare species because their conservation will contribute to the 2011-2020 targets in Global Strategy for Plant Conservation (www.cbd.int); targets which are still incomplete (Fenu et al. 2015). Thus, we can use rare species selected by Rabinowitz's and Gauthier's scheme to support decision-making, as rarity is considered a precursor of species extinction, and preventing plant extinction is the main goal of biodiversity conservation.

REFERENCES

Arens K (1958) O cerrado como vegetação oligotrófica. [The cerrado as oligotrophic vegetation.] *Bol da Fac Filos Ciências e Let Univ São Paulo Botânica* 57–77.

Arponen A (2012) Prioritizing species for conservation planning. *Biodivers Conserv* 21:875–893.

Bachelot B, Kobe RK (2013) Rare species advantage? Richness of damage types due to natural enemies increases with species abundance in a wet tropical forest. *J Ecol* 101:846–856.

Bachelot B, Kobe RK, Vriesendorp C (2015) Negative density-dependent mortality varies over time in a wet tropical forest, advantaging rare species, common species, or no species. *Oecologia* 179:853–861.

Bevill RL, Louda SM, Stanforth LM (1999) Protection from natural enemies in managing rare plant species. *Conserv Biol* 13:1323–1331.

Braun-Blanquet J (1951) *Pflanzensoziologie: grundzüge der vegetationskunde.* [Plant Sociology: Fundamentals of Vegetation Science.] Springer-Verlag.

Braun-Blanquet J (1979) *Fitosociología. Bases para el estudio de las comunidades vegetales.* [*Phytosociology. Bases for the study of plant communities.*].

Brasil. 2013. *Serviço Geológico do Brasil. Mapa de declividade em percentual do relevo brasileiro.* [Geological Survey of Brazil. Map of declivity in percentage of Brazilian relief.] Disponível em <http://www.cprm.gov.br/publique/Gestao-Territorial/Geodiversidade/ Mapa-de-Declividade-em-Percentual-do-Relevo-Brasileiro-3497.html>. Acessed 20 December 2017.

Brito D (2010) Overcoming the Linnean shortfall: data deficiency and biological survey priorities. *Basic Appl Ecol* 11:709–713.

Broennimann O, Vittoz P, Moser D, Guisan A (2005) Rarity types among plant species with high conservation priority in Switzerland. *Bot Helv* 115:95–108.

Brown J, Enright NJ, Miller BP (2003) Seed production and germination in two rare and three common co-occurring Acacia species from south-east Australia. *Austral Ecol* 28:271–280.

Caiafa AN, Martins FR (2010) Forms of rarity of tree species in the southern Brazilian Atlantic rainforest. *Biodivers Conserv* 19:2597–2618.

Caro TM, O'doherty G (1999) On the use of surrogate species in conservation biology. *Conserv Biol* 13:805–814.

Castro A, Martins FR, Tamashiro JY, Shepherd GJ (1999) How rich is the flora of Brazilian cerrados? *Ann Missouri Bot Gard* 192–224.

Ceballos G, Ehrlich PR, Barnosky AD, et al. (2015) Accelerated modern human–induced species losses: Entering the sixth mass extinction. *Sci Adv* 1:e1400253.

Cofre H, Marquet PA (1999) Conservation status, rarity, and geographic priorities for conservation of Chilean mammals: an assessment. *Biol Conserv* 88:53–68.

Condit R, Engelbrecht BMJ, Pino D, et al. (2013) Species distributions in response to individual soil nutrients and seasonal drought across a community of tropical trees. *Proc Natl Acad Sci* 110:5064–5068.

Corlett RT (2016) Plant diversity in a changing world: status, trends, and conservation needs. *Plant Divers* 38:10–16.

Coutinho LM (1978) O conceito de Cerrado [The concept of Cerrado]. *Rev Bras Bot* 1:17–23.

Coutinho LM (2006) O conceito de bioma. [The biome concept.] *Acta Bot Bras* 20:13–23.

Dajoz R (1971) *Précis d'écologie* [Accurate ecology.]

de Lima RAF, Mori DP, Pitta G, et al. (2015) How much do we know about the endangered Atlantic Forest? Reviewing nearly 70 years of information on tree community surveys. *Biodivers Conserv* 24:2135–2148.

Dirzo R, Young HS, Galetti M, et al. (2014) Defaunation in the Anthropocene. *Science* (80-) 345:401–406.

Dynesius M, Jansson R (2000) Evolutionary consequences of changes in species' geographical distributions driven by Milankovitch climate oscillations. *Proc Natl Acad Sci* 97:9115–9120.

Eaton MA, Gregory RD, Noble DG, et al. (2005) Regional IUCN red listing: the process as applied to birds in the United Kingdom. *Conserv Biol* 19:1557–1570.

Eisenlohr PV, Oliveira-Filho AT (2015) Obtenção e estrutura de metadados para trabalhos fitogeográficos de síntese e o banco de dados NeotropTree como estudo de caso. [Obtaining and structure of metadata for synthesis phytogeographic works and the NeotropTree database as a case study.] In: Eisenlohr PV, Felfili MJ, Melo MMRF, Andrade LA, Meira-Neto JAAM (2015) *Fitossociologia no Brasil: métodos e estudos de casos*. Editora UFV.

Elton CS (1927) *Animal ecology*. University of Chicago Press.

ESRI 2011. ArcGIS Desktop: Release 10. Redlands, CA: Environmental Systems Research Institute.

Feeley K (2015) Are we filling the data void? An assessment of the amount and extent of plant collection records and census data available for tropical South America. *PLoS One* 10:e0125629.

Feeley KJ, Silman MR (2011) The data void in modeling current and future distributions of tropical species. *Glob Chang Biol* 17:626–630.

Fenu G, Fois M, Cogoni D, et al. (2015) The Aichi Biodiversity Target 12 at regional level: an achievable goal? *Biodiversity* 16:120–135.

Fiedler PL (1995) Rarity in the California flora: new thoughts on old ideas. *Madrono* 127–141.

Field KG, Coddington J (1980) Rare plant species in Massachusetts. *Rhodora* 82:151–162.

Forzza RC et al. (2012) New Brazilian floristic list highlights conservation challenges. *BioScience* 62:39-45.

Françoso RD, Haidar RF, Machado RB (2016) Tree species of South America central savanna: endemism, marginal areas and the relationship with other biomes. *Acta Bot Brasilica* 30:78–86.

Game ET, Kareiva P, Possingham HP (2013) Six common mistakes in conservation priority setting. *Conserv Biol* 27:480–485.

Gaston KJ (1994) What is rarity? In: *Rarity*. Springer, pp 1–21.

Gaston KJ (2009) *Geographic range limits of species.*

Gaston KJ, Fuller RA (2009) The sizes of species' geographic ranges. *J Appl Ecol* 46:1–9.

Gauthier P, Debussche M, Thompson JD (2010) Regional priority setting for rare species based on a method combining three criteria. *Biol Conserv* 143:1501–1509.

Giehl ELH, Jarenkow JA (2012) Niche conservatism and the differences in species richness at the transition of tropical and subtropical climates in South America. *Ecography (Cop)* 35:933–943.

Gleason HA (1929) The significance of Raunkiaer's law of frequency. *Ecology* 10:406–408.

Goodland R, Pollard R (1973) The Brazilian cerrado vegetation: a fertility gradient. *J Ecol* 219–224.

Gotelli NJ, Colwell RK (2001) Quantifying biodiversity: procedures and pitfalls in the measurement and comparison of species richness. *Ecol Lett* 4:379–391.

Grinnell J (1917) The niche-relationships of the California Thrasher. *Auk* 34:427–433.

Groves CR, Jensen DB, Valutis LL, et al. (2002) Planning for Biodiversity Conservation: Putting Conservation Science into Practice: A seven-step framework for developing regional plans to conserve biological diversity, based upon principles of conservation biology and ecology, is being used extensively. *AIBS Bull* 52:499–512.

Guo Q (2003) Plant abundance: the measurement and relationship with seed size. *Oikos* 101:639–642.

Hartley S, Kunin WE (2003) Scale dependency of rarity, extinction risk, and conservation priority. *Conserv Biol* 17:1559–1570.

Haslouer SG, Eberle ME, Edds DR, et al. (2005) Current status of native fish species in Kansas. *Trans Kansas Acad Sci* 108:32–46.

Henderson PA, Magurran AE (2010) Linking species abundance distributions in numerical abundance and biomass through simple assumptions about community structure. *Proc R Soc London B Biol Sci* rspb20092189.

Holt RD, Barfield M (2009) Trophic interactions and range limits: the diverse roles of predation. *Proc R Soc London B Biol Sci* 276:1435–1442.

Hubbell SP (2013) Tropical rain forest conservation and the twin challenges of diversity and rarity. *Ecol Evol* 3:3263–3274.

Izco J (1998) Types of rarity of plant communities. *J Veg Sci* 9:641–646.

Klironomos JN (2002) Feedback with soil biota contributes to plant rarity and invasiveness in communities. *Nature* 417:67.

Knapp S, Salomón JM (2010) A simple method for assessing preliminary conservation status of plants at a national level: a case study using the ferns of El Salvador. *Oryx* 44:523–528.

Kricsfalusy V V, Trevisan N (2014) Prioritizing regionally rare plant species for conservation using herbarium data. *Biodivers Conserv* 23:39–61.

Kukkala AS, Moilanen A (2013) Core concepts of spatial prioritisation in systematic conservation planning. *Biol Rev* 88:443–464.

Kunin WE, Gaston KJ (1993) The biology of rarity: patterns, causes and consequences. *Trends Ecol Evol* 8:298–301.

Ladeira-Neto JF (2013) *Mapa de declividade em percentual do relevo brasileiro*. [Map of declivity in percentage of Brazilian relief.] Rio Janeiro CPRM.

Lambeck RJ (1997) Focal species: a multi-species umbrella for nature conservation. *Conserv Biol* 11:849–856.

Laurance WF, Useche DC (2009) Environmental synergisms and extinctions of tropical species. *Conserv Biol* 23:1427–1437.

Le Berre M, Noble V, Pires M, et al. (2018) Applying a hierarchisation method to a biodiversity hotspot: Challenges and perspectives in the South-Western Alps flora. *J Nat Conserv*.

Lennon JJ, Beale CM, Reid CL, et al. (2011) Are richness patterns of common and rare species equally well explained by environmental variables? *Ecography (Cop)* 34:529–539.

Lentini PE, Wintle BA (2015) Spatial conservation priorities are highly sensitive to choice of biodiversity surrogates and species distribution model type. *Ecography (Cop)* 38:1101–1111.

Lewandowski AS, Noss RF, Parsons DR (2010) The effectiveness of surrogate taxa for the representation of biodiversity. *Conserv Biol* 24:1367–1377.

Liancourt P, Callaway RM, Michalet R (2005) Stress tolerance and competitive-response ability determine the outcome of biotic interactions. *Ecology* 86:1611–1618.

Linklater WL (2003) Science and management in a conservation crisis: a case study with rhinoceros. *Conserv Biol* 17:968–975.

Liu Y, Shen Z, Wang Q, et al. (2017) Determinants of richness patterns differ between rare and common species: implications for Gesneriaceae conservation in China. *Divers Distrib* 23:235–246.

Lõhmus A (2015) Collective analyses on "red-listed species" may have limited value for conservation ecology. *Biodivers Conserv* 24:3151–3153.

Lopes AS, Cox FR (1977) Cerrado Vegetation in Brazil: An Edaphic Gradient1. *Agron J* 69:828–831.

Mace GM (2004) The role of taxonomy in species conservation. *Philos Trans R Soc London B Biol Sci* 359:711–719.

Maciel EA, Eisenlohr P V (2016) On the collective analysis of species: how can Red Lists and lists of regional priorities be combined to assist in decision-making? A reply to Lõhmus (2015). *Biodivers Conserv* 25:611–614.

Maciel EA, Oliveira-Filho AT, Eisenlohr P V (2016) Prioritizing rare tree species of the Cerrado-Amazon ecotone: warnings and insights emerging from a comprehensive transitional zone of South America. *Nat Conserv* 14:74–82.

Mangan SA, Schnitzer SA, Herre EA, et al. (2010) Negative plant–soil feedback predicts tree-species relative abundance in a tropical forest. *Nature* 466:752.

Margules CR, Pressey RL (2000) Systematic conservation planning. *Nature* 405:243–253.

Margules CR, Pressey RL, Williams PH (2002) Representing biodiversity: data and procedures for identifying priority areas for conservation. *J Biosci* 27:309–326.

McIntyre S (1992) Risks associated with the setting of conservation priorities from rare plant species lists. *Biol Conserv* 60:31–37.

Miranda S do C, Bustamante M, Palace M, et al. (2014) Regional variations in biomass distribution in Brazilian savanna woodland. *Biotropica* 46:125–138.

Mittermeier RA, Myers N, Thomsen JB, et al. (1998) Biodiversity hotspots and major tropical wilderness areas: approaches to setting conservation priorities. *Conserv Biol* 12:516–520.

Monteiro L, Machado N, Martins E, et al. (2017) *Conservation priorities for the threatened flora of mountaintop grasslands in Brazil*. Flora.

Mooney HA, Cleland EE (2001) The evolutionary impact of invasive species. *Proc Natl Acad Sci* 98:5446–5451.

Murray BR, Lepschi BJ (2004) Are locally rare species abundant elsewhere in their geographical range? *Austral Ecol* 29:287–293.

Myers N, Mittermeier RA, Mittermeier CG, et al. (2000) Biodiversity hotspots for conservation priorities. *Nature* 403:853–858.

Oliveira-Filho AT. 2017. NeoTropTree, Flora arbórea da Região Neotropical: Um banco de dados envolvendo biogeografia, diversidade e conservação. [NeoTropTree, Neotropical Tree Flora: A database involving biogeography, diversity and conservation.] Universidade Federal de Minas Gerais. (http://www.neotroptree.info).

Orme CDL, Davies RG, Burgess M. et al. (2005) Global hotspots of species richness are not congruent with endemism or threat. *Nature* 436:1016.

Pandit SN, Kolasa J, Cottenie K (2009) Contrasts between habitat generalists and specialists: an empirical extension to the basic metacommunity framework. *Ecology* 90:2253–2262.

Pärtel M, Helm A, Ingerpuu N, et al. (2004) Conservation of Northern European plant diversity: the correspondence with soil pH. *Biol Conserv* 120:525–531.

Pearman PB, Weber D (2007) Common species determine richness patterns in biodiversity indicator taxa. *Biol Conserv* 138:109–119.

Pereira HM, Ferrier S, Walters M, et al. (2013) Essential biodiversity variables. *Science* 339:277–278.

Pereira LC, Lombardi-Neto F (2004) *Avaliação da aptidão agrícola das terras: proposta metodológica.* [Assessment of agricultural suitability of land: methodological proposal.] Embrapa Meio Ambiente Doc.

Pilgrim ES, Crawley MJ, Dolphin K (2004) Patterns of rarity in the native British flora. *Biol Conserv* 120:161–170.

Pires JM, Prance GT (1977) The Amazon Forest: a natural heritage to be preserved. In: *Extinction is Forever.* Bronx, NY (USA). 1976.

Pitman NCA, Terborgh J, Silman MR, Nunez V (1999) Tree species distributions in an upper Amazonian forest. *Ecology* 80:2651–2661.

Possingham HP, Grantham H, Rondinini C (2007) How can you conserve species that haven't been found? *J Biogeogr* 34:758–759.

Prendergast JR, Quinn RM, Lawton JH, et al. (1993) Rare species, the coincidence of diversity hotspots and conservation strategies. *Nature* 365:335.

Price TD, Kirkpatrick M (2009) Evolutionarily stable range limits set by interspecific competition. *Proc R Soc London B Biol Sci rspb*-2008.

R Core Team (2017). R: *A language and environment for statistical computing*. R Foundation for Statistical Computing, Vienna, Austria. URL https://www.r-project.org/

Rabinowitz D (1981a) Seven forms of rarity. In: Synge H (ed) The *Biological Aspects of Rare Plant Conservation*. New York: Wiley, pp. 205-217.

Rabinowitz D (1978) Abundance and diaspore weight in rare and common prairie grasses. *Oecologia* 37:213–219.

Rabinowitz D (1981b) Seven forms of rarity. In "*The biological aspects of rare plant conservation*." (Ed. H Synge) pp. 205–217.

Rabinowitz D, Rapp JK, Dixon PM (1984) Competitive abilities of sparse grass species: means of persistence or cause of abundance. *Ecology* 65:1144–1154.

Ratter JA, Bridgewater S, Atkinson R, Ribeiro JF (1996) Analysis of the floristic composition of the Brazilian cerrado vegetation II: comparison of the woody vegetation of 98 areas. *Edinb J Bot* 53:153–180.

Ratter JA, Bridgewater S, Ribeiro JF (2003) Analysis of the floristic composition of the Brazilian cerrado vegetation III: comparison of the woody vegetation of 376 areas. *Edinburgh J Bot* 60:57–109.

Rey Benayas JM, Scheiner S, García Sánchez-Colomer M, Levassor C (1999) Commonness and rarity: theory and application of a new model to Mediterranean montane grasslands. *Conserv Ecol* 3.

Ribeiro JF, Walter BMT (2008) As principais fitofisionomias do bioma cerrado in: Sano, SM; Almeida, SP; Ribeiro, JF Cerrado: *Ecologia e flora*. Brasília: Embrapa Informação Tecnológica.

Rocap G, Larimer FW, Lamerdin J, et al. (2003) Genome divergence in two Prochlorococcus ecotypes reflects oceanic niche differentiation. *Nature* 424:1042.

Sætersdal M, Birks HJB (1997) A comparative ecological study of Norwegian mountain plants in relation to possible future climatic change. *J Biogeogr* 24:127–152.

Schmeller DS, Gruber B, Budrys E, et al. (2008) National responsibilities in European species conservation: a methodological review. *Conserv Biol* 22:593–601.

Scudeller VV, Martins FR, Shepherd GJ (2001) Distribution and abundance of arboreal species in the Atlantic ombrophilous dense forest in Southeastern Brazil. *Plant Ecol* 152:185–199.

Simon MF, Du Vall Hay J (2003) Comparison of a common and rare species of Mimosa (Mimosaceae) in Central Brazil. *Austral Ecol* 28:315–326.

Simon MF, Proença C (2000) Phytogeographic patterns of Mimosa (Mimosoideae, Leguminosae) in the Cerrado biome of Brazil: an indicator genus of high-altitude centers of endemism? *Biol Conserv* 96:279–296.

Soberón J (2007) Grinnellian and Eltonian niches and geographic distributions of species. *Ecol Lett* 10:1115–1123.

Sólymos P, Feher Z (2005) Conservation prioritization based on distribution of land snails in Hungary. *Conserv Biol* 19:1084–1094.

Soulé ME, Estes JA, Berger J, Del Rio CM (2003) Ecological effectiveness: conservation goals for interactive species. *Conserv Biol* 17:1238–1250.

Spitale D (2012) A comparative study of common and rare species in spring habitats. *Ecoscience* 19:80–88.

Synge H (1980) The biological aspects of rare plant conservation. In: Chichester etc.: John Wiley and Sons xxviii, 558p.-illus., maps. *En Proceedings of International Conference,* King's College, Cambridge. pp 14–19.

Tilman D, May RM, Lehman CL, Nowak MA (1994) Habitat destruction and the extinction debt. *Nature* 371:65.

Vera P, Sasa M, Encabo SI, et al. (2011) Land use and biodiversity congruences at local scale: applications to conservation strategies. *Biodivers Conserv* 20:1287–1317.

Villalobos F, Lira-Noriega A, Soberón J, Arita HT (2013) Range–diversity plots for conservation assessments: Using richness and rarity in priority setting. *Biol Conserv* 158:313–320.

Wamelink GWW, Goedhart PW, Frissel JY (2014) Why some plant species are rare. *PLoS One* 9:e102674.

Whittaker RH (1965) Dominance and diversity in land plant communities: numerical relations of species express the importance of competition in community function and evolution. *Science* 147:250–260.

Williams SE, Williams YM, VanDerWal J, et al. (2009) Ecological specialization and population size in a biodiversity hotspot: how rare species avoid extinction. *Proc Natl Acad Sci* 106:19737–19741.

Wills C, Condit R, Foster RB, Hubbell SP (1997) Strong density-and diversity-related effects help to maintain tree species diversity in a neotropical forest. *Proc Natl Acad Sci* 94:1252–1257.

Wilson K, Pressey RL, Newton A, et al. (2005) Measuring and incorporating vulnerability into conservation planning. *Environ Manage* 35:527–543.

Wolkovich EM, Cook BI, Davies TJ (2014) Progress towards an interdisciplinary science of plant phenology: building predictions across space, time and species diversity. *New Phytol* 201:1156–1162.

Xu H, Wu J, Liu Y, et al. (2008) Biodiversity congruence and conservation strategies: a national test. *AIBS Bull* 58:632–639.

Zappi DC et al. (2015). Growing knowledge: an overview of seed plant diversity in Brazil. *Rodriguésia* 66:1085-1113.

Zhao L, Li J, Liu H, Qin H (2016) Distribution, congruence, and hotspots of higher plants in China. *Sci Rep* 6:19080.

In: Forest Conservation
Editor: Pedro V. Eisenlohr

ISBN: 978-1-53614-559-5
© 2019 Nova Science Publishers, Inc.

Chapter 2

SUSTAINABLE MANAGEMENT OF BIODIVERSITY IN WOODY ECOSYSTEMS: BIOTECHNOLOGY AND BIOPROSPECTING OF NATIVE SPECIES FROM MONTE DESERT, PATAGONIA

Patricia Boeri[1,2,], Lucrecia Piñuel[1,2], Daniela Dalzotto[1,2], Daniel Barrio[1,2], Maite Romero Alves[3], Marianelén Cedrés Gazo[4] and Sandra Sharry[1,2,3,5]*

[1] CIT-RIO NEGRO Sede Atlántica, Universidad Nacional de Rio Negro (UNRN-CONICET), Viedma, Rio Negro, Argentina
[2] UIISA-Unidad Integrada para la Innovación del Sistema Agroalimentario de la Patagonia Norte. Viedma, Río Negro, Argentina
[3] LIMAD-Facultad de Ciencias Agrarias y Forestales-Universidad Nacional de La Plata, La Plata, Buenos Aires, Argentina

---

[*] Corresponding Author Email: pboeri@unrn.edu.ar.

[4]Subsecretaría de Recursos Forestales, Ministerio de Agricultura, Ganadería y Pesca de la Provincia de Río Negro, Viedma, Río Negro, Argentina
[5]CIC-PBA Comisión de Investigaciones Científicas-Buenos Aires, Argentina, La Plata, Buenos Aires, Argentina

INTRODUCTION

Dry-land surface (i.e., semi-arid and arid regions) covers approximately one-third of the world's land. During the last 50 years, there has been growing interest in the causes of desertification due to its impact on the global environment, economy and society. These dry-ecosystems provide a series of ecological services, which are essential for the sustainability of human life and the development of productive activities. The arid and semi-arid environments represent approximately 60% of the continental surface of Argentina and they are divided into three different phytogeographical provinces: Patagonia, Puna, and Monte. Particularly, this last area extends from 24°35´S in Salta Province to 44° 20´ S in Chubut Province; and from 69° 50´ W at the foot of the Andes to 62° 54´ W of the Atlantic coast as well as from 69° 50´ W. The Monte is a warm shrub desert, which has a particular biogeographical interest as it lies between the Neotropical and Antarctic regions (Roig-Juñent and Flores 2001).

Monte is dominated by arid and semi-arid to arid conditions. The average annual precipitation varies between 30-350 mm, and the relationship between precipitation and potential evapotranspiration ranges between 0.05 and 0.5, indicating a water deficit in the entire area (Rundel et al. 2007; Villagra et al. 2004). The southernmost region of the Monte is part of extra-Andean Argentinian Patagonia, an area of treeless semi-arid grass and shrub steppes that has been overgrazed for over a century. These semi-arid regions are very susceptible to environmental impacts and they are exposed to an easy destruction of their resources, if they do not carry out rational management of them (Fiori and Zalba 2000). These habitats are affected by different disturbances, such as the depletion of the cover of woody vegetation due to the felling of trees for fuel, forest management practices that alter the fire regime and excessive grazing, especially during periods of drought (Milesi et al. 2002). On the other hand, due to the exposition of its meristems, fire mainly affects the shrubs (Kröpfl et al. 2007).

These arid and semi-arid biomes of Argentina are undergoing rapid habitat conversion (Ojeda et al. 1998). The anthropic activity, and fundamentally the change in the use of the land, have triggered a set of environmental problems that directly or indirectly affect biodiversity. The new systems of agricultural production added to an extractive culture have caused an increase of the genetic erosion at an alarming rate. In the past few years, Argentina has faced one of the processes of transformation of its largest native forests in the history (Montenegro et al. 2005). In the last century, this country lost 70% of its native forests. This process does not stop (La Rocca, United Nations Convention to Combat Desertification, 2009). Furthermore, the Patagonian region has not been exempt from this situation, being currently subject to grazing and mining activities that result in a moderate to severe desertification (Mazzonia and Vazquez 2009). There, 80% of the surface today show signs of deterioration and desertification (Ravelo et al. 2011) and this affects their ability to generate goods and services.

In the extra-Andean Patagonia, overgrazing, deforestation and unsustainable agriculture are combined with natural erosion to carry ecosystems to critical points of deterioration (Ravelo et al. 2011). The 58% of its entire surface were classified as irreversible for the development of agricultural and livestock activities (Mazzonia and Vazquez 2009). Deforestation depletes biodiversity by destroying habitat, by fragmentation of areas and by interfering with plant reproduction. This process occurs by pulses associated with favorable moments for agricultural expansion, by either the prices of agricultural products, technological changes or socio-political context (Montenegro et al. 2005). Effects of this degradation on ecosystems and biodiversity are long lasting and devastating.

An alternative to restore this ecosystem, as well as other similarly degraded arid environments, is reforestation with native species obtained in nursery gardens (Maestre et al. 2011).

GOODS AND SERVICES PROVIDED BY DRY FOREST ECOSYSTEMS

The geographic extensions of Monte phytogeography regions/sub-regions are relative and change dynamically over the years depending on changes in the macroclimate. The forest area in these regions/sub regions was estimated by FAO and indicated that the Monte region has a native forest cover of approximately 36,000 ha and an area of forest plantations of less than 400 ha (FRA 2000). Vegetation in this dry forest is heterogeneously distributed resulting in a two-phase mosaic pattern formed by areas with high plant cover dominated by shrubs separated by areas of scattered vegetation or bare soil. Woody vegetation appears to have an important role in the origin, dynamics, and maintenance of spotted vegetation of the Monte (Bertiller et al. 2009). The dominant vegetation corresponds to a shrub steppe, characterized by the presence of *Larrea* genus (Jarilla), *Condalia microphylla* Cav. (Piquillin), *Monttea aphylla* (Miers.) Hauman (Mata de sebo), *Geoffroea decorticans* (Gill ex Hook et. Arn.) Burkart. (Chañar), *Prosopis flexuosa D.C* (Algarrobillo) and *Prosopis alpataco Phil.* (Alpataco).

These arid-ecosystems are particularly susceptible to climate change and desertification processes (Peters et al. 2012) and although they provide a wide range of goods and services, they are not as recognized as other terrestrial ecosystems on the planet. Arid and semi-arid areas cover 41% of the world's land surface. More than 38% of the human population needs ecosystem goods and services, but these depend largely on rain and soil fertility. Argentinian Monte provides a long list of goods and services related to biodiversity, including the prevention of soil erosion, pest control, pollination, drinking water, and food, mitigation of climate change, energy and water supply, fodder and resilience and stability of the local ecosystem (MEA 2005). Thus, there is need of conservation of these ecosystems.

The diversity and productivity of arid lands depend on the interaction between environmental limitations and the complex of adaptive characteristics that allow organisms to maintain their water and energy balance (Villagra et al. 2011). The life strategy of plant species defines the relative importance of the stages that make up their life cycles, as well as the biotic and abiotic factors that regulate them (Sagar and Mortimer 1976). In the same environment, you can find plants whose predominant propagation strategy is clonal reproduction, and others that are reproduced by seeds, which have different restrictions on germination. Among the latter, there are also species with different requirements for their perpetuation, which develop individual or combined strategies for survival under stress conditions (Gutterman 1994). The reproduction by seeds and vegetative multiplication represent the extremes of a continuum of possible regeneration alternatives, whose knowledge facilitates the implementation of conservation and sustainable management plans. Preserving plant diversity will be particularly important to maintain the quantity and quality of services provided by the ecosystems found in arid lands.

Ecological restoration of trees is often constrained by limited knowledge of the biology, propagation and management requirements of individual species. Globally, mixtures of native tree species are increasingly used to restore disturbed and degraded areas (McNamara et al. 2006; Shono et al. 2007; Raman et al. 2009). However, restoration programs rarely incorporate lesser-known species or consider as a starting point the study of the biology, propagation and management of woody species. The integration of this knowledge is vital for region-specific biodiversity conservation and restoration plans because it is a key factor for the recovery of these species within their natural habitat.

On the other hand, information on the potential of some native and exotic species to be used for industrial purposes in Patagonia is limited (Ravetta and Soriano 1998). While increasing plant biodiversity improves the capacity of ecosystems to provide multiple functions, drylands are widely used for food production. Many of our main food crops, such as wheat, barley, sorghum and millet, are originated in the drylands (White and Nackoney 2003).

In order to address societal challenges, like climate change, natural resource scarcity, and unsustainable consumption patterns, a transformative change is needed that involves long-term approaches and interactions at all levels of society. The concept of bioeconomy represents an opportunity to tackle these challenges and create the transformations needed in socio-technical systems (Van der Meulen et al. 2012). Currently, knowledge about the spread of the species that are part of the dry forest ecosystems and the use of these as a dendroenergy and food resource is scarce. Biotechnological techniques and bioprospecting strategies increasingly influence regional bioeconomies and people's lives, which is why it is necessary to develop integrated biotechnological conservation and bioprospection programs that optimize the propagation of the species involved and generate bioprospecting methodologies tending to know their potential nutritional and medicinal value.

Forest conservation is a national problem, so it must be approached with perfect coordination between the forests departments and other institutions involved. Research in the field of forest resources and deforestation is of vital importance.

In this sense, in order to take the first steps of integration between the scientific knowledge and contribute to the community about different biotechnological methodologies, a collaborative work was carried out between the Ministry of Agriculture, Livestock and Fisheries and the National University of Río Negro. We ask ourselves about: how can we expect the community to value and conserve genetic resources without having the necessary information about the native flora? How to generate a sustainable use of resources if the potential they possess is not known? How do they intend to conserve without knowing the reproductive strategies of each species and their adaptations to arid and semi-arid environments? To answer these questions, we conducted a series of experiments to study seed germination, seedling growth performance and early survival of seedlings of woody species native to arid environments. Our objective was to improve the basic knowledge of propagation techniques for these species in nurseries

Sustainable Management of Biodiversity in Woody Ecosystems 47

and to identify suitable species for the production of nurseries in restoration projects. Initially, the productive activity was based on the vegetative propagation of these species of arid environments. In this way, we were able to establish the first ex situ conservation methodologies of native and exotic species and have a large number of native shrubs to recover areas degraded by extractive activities and repopulate those affected by clearing, desertification and forest fires. The knowledge generated allowed to make available to the provincial community non-wood forest products and resources that promote productive development and environmental protection. Finally, the project allowed the creation of a basic laboratory of Plant Biotechnology to incorporate new methodologies to the current techniques of propagation and updating of the Viedma Forest Nursery.

OVERVIEW OF *EX SITU* CONSERVATION AND MANAGEMENT OF GENETIC RESOURCE FROM MONTE DRY FOREST

Biological diversity has already been threatened by habitat destruction and other human-caused stresses, and climate change makes its conservation even more of a challenge (Jackson and Kennedy 2009). Furthermore, in the arid ecosystem from Patagonia Monte, deforestation and the degradation of ecosystems have caused the destruction of the environment and the consequent loss of regional biodiversity.

Biodiversity conservation should be included in the land use planning and resource management strategies. It is important to emphasize those traditional methods of *ex situ* and *in situ* conservation must provide complementary means for the management of plant genetic resources (Benson 2002).

The aim of *ex situ* storage for conserving woody species is to maintain the initial genetic and physiological quality of the germplasm until it is used or regenerated. In recent years, the strengthening of environmental policies and programs for the recovery and conservation of ecosystems has promoted an increase in the demand for seeds of native species (Carvalho et al. 2006). Nevertheless, complementary and integrated *in situ* and *ex situ* conservation strategies are necessary (CBD 1992). However, among the semi-arid species that are under human pressure, only a few dozen populations or even individuals are available and conservation measures by *ex situ* collections are likely to be critical for the future of the species.

The biotechnological techniques provide new approaches to overcome the problems of plant under unfavorable environments, biotic stresses and in the field of plant germplasm.

In this sense, different *ex situ* conservation methodologies of forest trees and multiplication technologies can be used. There are a wide range of biotechnological methods, which can help plant conservation programs, such as tissue culture techniques, molecular genomes analysis, immunological diagnostics and cryoconservation protocols. Biotechnology has a major role in all aspects of plants genetic resources such as collection, characterization, disease indexing, propagation, patenting, storage, documentation and exchange of plant resources (Benson 2002).

When it comes to endanger or threatened plants, the use of *in vitro* tissue culture techniques can be of great interest for germplasm collection, storage and multiplication of recalcitrant and vegetatively propagated species (Engelmann 1991). Several studies have carried out the recapture of endangered or threatened species of the world through in *vitro* culture such us: *Betula uber* [Ashe] Fern. (Vijayakumar et al. 1990), *Physoplexis comosa* and *Primula glaucescens* (Cerabolini et al. 2004), *Helianthemum bystropogophyllum* (López et al. 2004), *Drosophyllum lusitanicum* (Goncalves and Romano 2005), *Helianthemum inaguae* (López et al. 2006), *Pinguicula lusitanica* (Goncalves et al. 2008), *Huernia hystrix* (Amoo et al. 2009), *Tuberaria major* (Goncalves et al. 2010), *Phytolacca tetrámera* (Cordal et al. 2014), *Rosa hybrida* L. cv. Helmut Schmidt (Mubbarakh et al. 2014) and *Gomortega keule* (Mol.) Baillon. (Calderon-Baltierra et al. 1993).

Fortunately, during the last decade, efforts have been devoted in Argentina to protect natural resources and promote a sustainable use of wildlife, but dry forests are still experiencing the increasing pressure of

Sustainable Management of Biodiversity in Woody Ecosystems 49

human activities, which are leading to a rapid deforestation and a loss of ecosystem services (Vilela et al. 2009).

In environments such as the arid ecosystems from Patagonia Argentina, where nutrient and water availability is very limited, plants have developed a wide variety of metabolites because of both selective pressure and resource availability (Vilela et al. 2011). Many of these adaptations are common in the different desert regions of the world and include phenological adjustments to the availability of water, such as morphophysiological characteristics that reduce the loss of water or increase its acquisition. In addition, plants use other mechanisms to the regulation of thermal stress and rapid response to pulses of precipitation such as germinating, growing, and producing large quantities of seed and different latency strategies (D'Antoni et al. 1977; Whitford 2002; Holmgren et al. 2006).

Many native shrubs from the arid region of Argentina are extracted systematically as dendroenergetic resources, mainly *Prosopis* spp., *Geoffroea decorticans* (chañar) and *Condalia microphylla* (piquillin) (Coirini and Karlin 2011). Moreover, there are other species that are used as non-wood forest products–either as food and food additives (edible seeds, fungi, fruits, fibers, spices and condiments, flavorings, wildlife, used for buildings, furniture, clothing or utensils) or as resins, gums, plant and animal products used for medicinal, cosmetic or cultural purposes (Ministry of Science, Technology and Productive Innovation). In this respect, many of those species are threatened or in danger of extinction such as *Acantholippia seriphioides* (tomillo) (Elechosa et al. 2009; Gonzalez et al. 2016). The World Conservation and Monitoring Centre estimates that more than 8000 tree species of the world are endangered (WCMC 2001). Therefore, thousands of tree species will depend on conservation outside protected areas: in managed forests, agricultural landscapes, or *ex situ* conditions. Furthermore, the Convention on Biological Diversity (CBD 1992) provides a legal framework for accessing, conserving and using biodiversity in a fair and equitable manner.

Our studies allowed us to optimize the germination processes of some native shrubs and obtain a massive production of native shrub seedlings, necessary for the successful restoration of the arid and semi-arid regions of

Patagonia. In addition, the knowledge of these methodologies allowed the introduction of juvenile material of woody species under *in vitro* conditions.

Simple sexual germination protocols were adjusted for native species addressed in this chapter, for example, *Prosopis alpataco, Prosopis caldenia* and *Geoffroea decorticans*. In these species, the chemical pre-treatments with sulfuric acid accelerated the process of germination without generating any damage in the emerging plants. On the other hand, *Acantholippia seriphioides* and *Condalia microphylla* showed low percentages of germination, less than 50%, even when different physical and chemical scarification were applied.

When juvenile explants of these species were introduced under in vitro conditions, different morphogenetic processes were achieved (organogenesis and embryogenesis) using Murashige and Skoog medium (Murashige and Skoog 1962). Particularly for *Prosopis alpataco*, shoots were obtained via direct organogenesis and roots by indirect organogenesis (Boeri 2016). On the other hand, somatic embryogenesis and whole plant regeneration was achieved in calli cultures derived from cotyledons of *in vitro* seedlings (Boeri and Sharry 2016). Since *Prosopis caldenia* only reproduces by seeds (Peláez et al. 1992), the study of this species in laboratory conditions is of great interest given the importance of the genus *Prosopis* for the vast arid and semi-arid zone of the country. So, roots were obtained by via direct and indirect organogenesis from *P. caldenia* (Cedres Gazo et al. 2015). For *Geoffroea decorticans* the rooting in MS medium was adjusted with different concentrations of acetic acid naphthalene (ANA) and the rooting percentage and the number of roots obtained increased along with the concentration of ANA (Dalzotto et al. 2017). Last, seed disinfection from *Acantholippia seriphioides* was adjusted and seedlings were obtained in vitro. When these were maintained in MS medium (without growth regulators) the mass production of shoots, branched stems and leaves was observed, due to the loss of apical dominance (Boeri et al. 2017).

Although there are limitations of the *in vitro* culture when it comes to woody species, the selected species have generated different types of morphogenic responses (Larroque 2012). Of the native shrubs evaluated, only *C. microphylla* presented alterations such as hyperhydricity and phenolic oxidation under *in vitro* conditions and high levels of endophytic contamination.

The plants obtained from sexual propagation and under *in vitro* conditions were kept under greenhouse conditions in the Forest Resources Secretariat (Ministry of Agriculture, Livestock and Fisheries of the Río Negro Province) to be used by this institution in the reforestation programs of the province.

Our studies showed that there is a morphogenetic response in low concentrations or even in the absence of growth regulators. Given that, plants grown under stress conditions produce secondary metabolites as defense mechanisms; these responses could be due to the presence of endogenous hormones in these species. For example, plants can synthesize Abscisic acid (ABA) in mature roots and leaves; particularly in response to water stress while in the seeds the ABA affects the induction and maintenance of some aspects of latency (Davies 2010).

The development and implementation of somatic embryogenic and other vegetative propagation methodologies do not only allow the *ex situ* conservation but also massive production of native species to foster ecosystem restoration and mitigate climate change. The long-term conservation of embryogenic cell lines can be a valuable tool for the long-term storage of plant germplasm by cryopreservation (İzgü and Mendi 2017). Our results on the development and implementation of *in vitro* tissue cultures and other propagation methodologies will not only allow incorporating these species into *ex situ* and *in situ* conservation programs of regional genetic diversity (Dalzotto et al. 2017) but also their massive propagation to restore degraded ecosystems and mitigate climate change (Boeri et al. 2013; 2015).

Developing countries are also applying biotechnology to forestry, with a number of commercialized applications. Micropropagation and biotechnology-based plant breeding account for 51% and 33% of the non-GM research projects, respectively, while the rest of the forestry research projects is on diagnostics or biobased pesticides and fertilizers (Oborne 2010). The application of biotechnology to primary production (includes all living natural resources, such as forests) could result in an emerging "bioeconomy" where biotechnology contributes to a significant share of economic output.

THE STATE OF BIOPROSPECTING IN DRY FOREST FROM MONTE ARGENTINIAN SPECIES

Historically, native forest are undervalued in relation to exotic plantations; so, management and policy programs focus on promoting their conservation, and this undoubtedly makes them more vulnerable. Lack of economic value of forest ecosystems and their linkages with human wellbeing restrict decision-makers in recognizing their significance (Rausser and Small 2000). The economic potential of biodiversity depends largely on its value. The development of new bioproducts and industries using resources from the flora and fauna, microorganisms and other biological resources is an objective for the valorization of biodiversity (Quezada et al. 2005). Phytogenetics resources are inputs for high-value-added activities based on knowledge. When genetic materials are candidate sources for new products, we can establish a starting point to start a bioprospecting study. In particular, it is essential to perform an analysis that identifies the conditions under which bioprospecting can create effective financial incentives based on the market for the conservation of biodiversity. Researchers of natural products are moving to organize useful ecological and taxonomic indicators in complete models in order to provide a systematic guide for the drug discovery process (Dreyfuss and Chapela 1994).

The Millennium Ecosystem Assessment (MEA 2005) allowed us to review ecosystem products (instead of services) and highlight the search for them in the context of bioprospecting. In this regard, UNEP, UN-FAO and the World Bank have encouraged the expansion the concept of bioprospecting towards the search for other products besides pharmaceutical, such as crop pollinators, biological control agents, organisms for biological control, resources phytogenetics for agriculture and

Sustainable Management of Biodiversity in Woody Ecosystems 53

forestry, species for aquaculture, bioremediation, ecological restoration and biological mining. They also have included agencies that offer new materials and designs with applications in engineering, manufacturing and construction. With this, the MEA has placed biodiversity at the center of industrial development. The resource for all bioprospecting is biodiversity (Beattie et al. 2011).

The forests are also livelihood sources for a large number of communities that depend upon them (Verma et al. 2017), for example, *Prosopis* species fruits constitute a food source for human and animal of Monte desert (Arenas 2003). However, despite this, native forests have always been undervalued compared to those with implanted species. This situation could be reversed if the social and productive value of the native forests is given an identity. Therefore, it is very important to generate knowledge about the potential and current properties of our resources through bioprospecting.

Some geographic regions have particular prominence as sources of valuable research leads. The biodiversity of dryland ecosystems offers an immense source of genes and molecules that are a potential source of new bioproducts obtained by biotechnology. In these environments, natural selection favors the presence of species with a wide array of complex chemical means of defense (Coley et al. 1985), which translates into the synthesis of secondary metabolites (Vilela et al. 2011). The regional phytogenetic diversity provides the possibility to obtain products derived from bioprospecting.

Our workgroup studied the chemical and nutritional composition of the fruits of *Prosopis alpataco* from the Northern Patagonia Argentina, a native species of the arid zone, suggesting that integral flour could be used as a dietary supplement for the food industry, either in the preparation of food for human or animal consumption (Boeri et al. 2017). The plant genus *Prosopis*, including *P. alpataco*, has been described; important applications in medicinal products for human use as well as in veterinary medicine include antidiabetic, anti-inflammatory, anticancer, and antimicrobial activities (Henciya et al. 2017; Castro 2017). However, biological activity of *P. caldenia* from Monte Patagonian has not yet been described.

In view of these advantageous functional-food properties and the opportunity to exploit this Patagonian natural resource, *Condalia microphylla* (piquillin) was also evaluated. Local communities consume their drupes as fresh fruits, so we also quantified the nutritional value of the fruit and the phenolic compounds present to determine the functional antioxidant properties *in vitro* and *in vivo* (Boeri et al. 2018). Moreover, the presence of bioactives proteins with potential antitumoral activity was also evaluated (Piñuel et al. 2015). There is a lack of scientific information about their activity and/or chemical pattern of this plant. Cespedes et al. (2013) showed that extract from *Condalia microphylla* has insect growth inhibitory activity. On the other hand, Carpinella et al. (2009) found that extract of this plant has a 47% of acetylcholinesterase inhibitory activity, bioactive compounds for the treatment of neurological disorders.

Other species from the Monte, such as *Prosopis caldenia, Geoffroea decorticans* and *Acantholippia seriphioides,* at current, are being investigated regarding their biological and nutritional properties. Among these species, *Geoffroea decorticans* from North Argentinian has been most studied (Costamagna et al. 2013, 2016; Reynoso et al. 2013, 2016). However, there is scientific evidence that in some species indistinguishable from a botanical point of view, the presence of chemotypes has been reported, according to the atmospheric and edaphic conditions in which the individuals grow (Soria et al. 2008). In that respect, biological and nutritional properties of *Geoffroea decorticans* from Patagonian Argentinian had not yet been studied. Moreover, biological activities of *A. seriphioides* have been evaluated in several regions of the central, northwest and south regions of the country; however, knowledge is only related to this essential oil (Gonzalez et al. 2016).

The knowledge of nutritional and functional properties of native plants could stimulate the preservation of forests including sustainable development and management for commercial purposes, contributing to regional development in arid zones. Much of the controversy has to do with issues of ownership of plant genetic materials and access to legislation concerning material and intellectual property. Summarizing, the problem is of assessing the value of a genetic resource and agreeing on what constitutes a fair deal.

CONCLUSION

Ignoring the importance of biodiversity as a natural heritage is one of the main threats to global biodiversity. At a global level, perspectives are being analyzed to advance towards an economy based on biodiversity and ecosystem services.

The dry forest ecosystems of Monte, Argentina, are important livelihoods for the sustenance of rural populations. However, the lack of knowledge about their biodiversity causes the undervaluation of these environments, within other causes. This situation limits decision-making and management, necessary to recognize its importance. The dryland ecosystem is very susceptible to environmental impacts and is exposed to the loss of its resources, so it is necessary to make a rational management of them. In this context, new approaches like biotechnology and bioprospection play a strategic role since they allow the generation of technological developments based on our genetic resources. Bioprospection provides the necessary knowledge to find new products and characterize potential sources of food, medicines and aromatic-medicinal plants. These properties give the native flora an additional value and directly impact as benefits for people. On the other hand, propagation, characterization and conservation programs allow guaranteeing the long-term permanence of these plant genetic resources.

Our results suggest that arid and semi-arid regions have a great potential for the development of these technologies, in particular, biotechnology and bioprospecting for the promotion of biodiversity conservation and sustainable development. The integration of these technologies and science will be able to develop more sustainable production patterns from the economic, social and environmental point of view and favor regional and local developments. For this, it is necessary to promote research projects that allow a better evaluation and understanding of the economic and social value of biological diversity, which could help to design appropriate policy measures to balance conservation and economic development in the region. Thus, the creation of new scientific frameworks derived from traditional disciplines will allow the design of adequate policy measures to balance the conservation and economic development of a region. In addition, this strategy will allow generating sustainable production systems and re-positioning the Patagonia region towards a green economy. Thus, it will be possible to promote alternatives that facilitate the generation of employment in regional communities, especially in rural sectors. On the other hand, efforts can be intensified to use our biodiversity in a sustainable and economic way in order to support the livelihoods of the local population, including present and future generations.

This requires the integration of the different parties involved, where the government plays an active role, with a broader view of bioprospecting, understanding it not only as a mean to generate resources for the community, but also to sustain the necessary infrastructure for the consolidation of knowledge production chains (for example, universities, research institutions).

One of the biggest challenges focuses on data management. Confidential data is a significant aspect of the culture of bioprospecting. Although there are some treaties and international laws tending to prevent biopiracy, they have certain fragility about the generation of legal frameworks to protect biodiversity and generate a fair and equitable distribution of the benefits derived from it.

In this sense, bioprospecting programs are increasingly involved in issues of equity and ownership of intellectual property.

REFERENCES

Amoo, SO; Finnie JF; & Van Staden J. (2009). In vitro propagation of Huernia hystrix: an endangered medicinal and ornamental succulent. *Plant Cell Tissue and Organ Culture 96: 273–278.*

Arenas, P. (2003). Ethnography and food among the toba - ñachilamolekek y Wichí - Lhukútas of the Central Chaco (Argentina). Ed. Pastor Arenas. Buenos Aires, Argentina.

Beattie, AJ; Hay, M; Magnusson, B; de Nys, R; Smeathers, J; & Vincent, JFV. (2011). Ecology and bioprospecting. *Austral Ecology 36(3):* 341–356.

Benson, E. (Ed.). (2002). *Plant conservation biotechnology.* CRC Press 2002-309 pp.

Bertiller, MB; Marone, L; Baldi, R; & Ares, JO. (2009). Biological interactions at different spatial scales in the Monte desert of Argentina. *Journal of Arid Environments 73(2):* 212-221.

Boeri, P. (2016). Chemical bioprospecting and propagation of native plants of the Patagonian monte as strategies of conservation and sustainable use. Doctoral dissertation, Facultad de Ciencias Agrarias y Forestales, Universidad Nacional de La Plata. 387pp.

Boeri, P., & Sharry, S. (2016). *Somatic embryogenesis of P. alpataco. Stepwise protocols for somatic Embryogenesis of woody plants. Argentina.* LA PLATA. IUFRO Unit 2.09.02 on Development and application of vegetative propagation technologies in plantation forestry to cope with a changing climate and environment. Fac Cs Agrarias y Forestales, UNLP and IUFRO. Available from: https://www.iufro. org/page-not-found/.

Boeri, P; Cedrés Gazo, M; Sharry, S; Baffoni, JC; & Barrio, D. (2015). Planting future, recovery of forest environments with native species in northern Patagonia, Argentina. *XIV World Forestry Congress.* FAO. Durban, Sudáfrica. Available from: http://foris.fao.org/wfc2015/ api/file/5545441de52d79267e89a3da/contents/6109937c-b8dc47b3-91f9-fa1f46ec9f3a.pdf.

Boeri, P; Piñuel, L; Sharry, S; & Barrio, D. (2017). Nutritional characterization of the integral flour of algarrobo (*Prosopis alpataco*) of northern Patagonia Argentina. *Revista de la Facultad de Agronomía, La Plata 116(1)*, 129-140.

Boeri, P; Piñuel, L; Sharry, S; & Barrio, D. (2018). Chemical and Biological Characterization from *Condalia microphylla* Fruits, a Native Species of

Patagonia Argentina. In press. *Journal of agricultural Science and technology* Volume 7, Number 6B.

Boeri, P; Romero Alves, M; Dalzotto, D; & Sharry, S. (2017). Preliminary results of tissue culture of "tomillo silvestre" *Acantholippia seriphioides* (A. Gray) Moldenke. *REDBIO 2017. Bahía Blanca. Argentina*. Sección Biotecnología Vegetal (Bv20) Available from: http://www.redbio argentina.org.ar/simposio-2017-bahia-blanca/.

Boeri, P; Sharry, S; Baffoni, JC; & Barrio, D. (2013). Multiplication of Native Species and sustainable management of the Rionegrino xerophytic forest. *Congreso Forestal Argentino y Latinoamericano AFOA* N°229 ISSN 1669-6786.

Calderon-Baltierra, X; Perez, F; & Rotella, A. (1993). *In vitro* propagation of a Chilean endangered plant species: *Gomortega keule* (Mol.) Baillon (Magnoliopsidae, Gomortegaceae). *Bosque 14*, 23-28.

Carpinella, MC; Andrione, DG; Ruiz, G; & Palacios, SM. (2010). Screening for acetylcholinesterase inhibitory activity in plant extracts from Argentina. *Phytotherapy research 24(2):* 259-263.

Carvalho, RT; Da Silva, EA; & Davide, AC, (2006), Storage behaviour of forest sedes. *Revista Brasileira de Sementes 28(2):* 15-25.

Castro, D; Urzúa, J; Rodriguez-Malebran, M; Inostroza-Blancheteau, C; & Ibáñez, C. (2017). Woody leguminous trees: New uses for sustainable development of drylands. *Journal of Sustainable Forestry 36(8):* 764-786.

CBD - Convention on Biological Diversity CBD. (1992). *Global Strategy for Plant Conservation.* 30p. Available from: http://69.90.183.227/doc/meetings/cop/cop-07/official/cop-07-10-es.pdf.

Cedres Gazo, M; Boeri, P; Dalzotto, D; Barrio, D; Baffoni, JC; & Sharry, S. (2015). Organogenesis in vitro of *Prosopis caldenia. REDBIO 2015*. San Miguel de Tucumán, Argentina. Available from: http://redbioargentina.org.ar/simposio2015/resumenes/presentados/BV 43.pdf.

Cerabolini, B; de Andreis, R; Ceriani, RM; Pierce, S; & Raimondi, B. (2004). Seed germination and conservation of endangered species from the Italian Alps: *Physoplexis comosa* and *Primula glaucescens. Biological Conservation 117:* 351–356.

Cespedes, CL; Molina, SC; Muñoz, E; Lamilla, C; Alarcon, J; Palacios, SM; Carpinella, M; & Avila, JG. (2013). The insecticidal, molting disruption and insect growth inhibitory activity of extracts from *Condalia microphylla* Cav. (Rhamnaceae). *Industrial crops and products 42*, 78-86.

Coirini, R.; & Karlin, M. (2011). Technical report in the framework of the consultancy: Manual of Good Practices and Models of Sustainable Production: Models of Sustainable Production Espinal Ecorregion. Forestry and land use, 17p. Freiburg. Available from: Available from: https://www.argentina.gob.ar/ambientearchivos/web/CompBosNatBio/file/Espinal%20-%20Introduccion.pdf.

Coley, PD; Bryant, JP; & Chapin, FS. (1985). Resource availability and plant antiherbivore defense. *Science 230*(4728): 895-899.

Cordal, BMA; Adema, V; Briones, B; Villarreal, M; Panarisi, H; Abedini, W; & Sharry, S. (2014). Induction of somatic embryogenesis in *Phytolacca tetramera*, medicinal species of Argentina. *Emirates Journal of Food and Agriculture; Al-Ain 26(6):* 552-557.

Costamagna, MS; Ordoñez, RM; Zampini, IC; Sayago, JE; & Isla, MI. (2013). Nutritional and antioxidant properties of *Geoffroea decorticans* an Argentinean fruit, and derived products (flour, arrope, decoction and hydroalcoholic beverage) *Food Research International 54*: 160-168.

Costamagna, MS; Zampini, IC; Alberto, MR; Cuello, S; Torres, S; Pérez, J; Quispe, C; Schmeda-Hirschmann, G; & Isla, MI. (2016). Polyphenols rich fraction from *Geoffroea decorticans* fruits flour affects key enzymes involved in metabolic syndrome, oxidative stress and inflammatory process. *Food Chemistry 190*: 392-402.

Dalzotto, D; Boeri, P; Romero Alves, M; Cedres Gazo, M; López Dumrauf, I; & Sharry, S. (2017). Conservation of biodiversity through micropropagation strategies for forest species. *V National Congress of Conservation of Biodiversity - 1st ed. - CABA: Felix de Azara Natural History Foundation*, pag 200. Available from: http://fundacion azara.org.ar/img/otras-publicaciones/Libro-de-resumenes-VCNCB.pdf.

D'Antoni, HL; & Solbrig, OT. (1977). Algarrobos in South American Cultures: Past and Pre-sent. In: *Mesquite, its Biology in Two Desert Ecosystems*. Simpson, BB (Ed.), Dowden, Hutchinson & Ross Stroudsburg, Pennsylvania. 189-199 pp.

Davies, PJ. (2010). The Plant Hormones: Their Nature, Occurrence, and Functions. In: Davies P.J. (Eds) *Plant Hormones*. Springer, Dordrecht 1-15 pp.

Dreyfuss, MM; & Chapela, IH. (1994). Potential of fungi in the discovery of novel, low-molecular weight pharmaceuticals. In: *Discovery of Novel Natural Products with Therapeutic Potential* 49-80 pp.

Elechosa, MA; Aguirre, E; Bandoni, AL; Di Leo Lira PMR; Fernández, EA; Heit, C; Juárez, MA; López, S; Martínez, A; Martínez, E; Marino, AM; Molina, AC; van Baren, CM; & Viturro, CI. (2009). *Handbook of sustainable collection of native aromatic plants from the central and northwestern region of Argentina*. Ed. INTA, Bs As, Argentina 63 pp.

Engelmann, F. (1991). *In vitro* conservation of tropical plant germplasm - a review. *Euphytica 57*: 227-243.

Fiori, SM; & Zalba, SM. (Eds.). (2000). *Auca Mahuida Provincial Reserve Management Plan (Neuquén)*. Vol. I - Regional Diagnosis. Secretaría de Estado COPADE- CFI. Neuquén.

FRA- Forest Resources Assessment - FAO (2000) forest cover assessment In: *The argentinean regions of monte and espinal. Forest Resources Assessment Programme*. Paper 48. Available from: http://www.fao.org /docrep/006/ad673e/ad673e01.htm#TopOfPage.

Goncalves, S; & Romano, A. (2005). Micropropagation of *Drosophyllum lusitanicum* (L.) Link. an endangered West Mediterranean endemic insectivorous plant. *Biodiversity and Conservation 14*: 1071–1081.

Goncalves, S; Escapa, AL; Grevenstuk, T; & Romano, A. (2008). An efficient in vitro propagation protocol for *Pinguicula lusitanica*, a rare insectivorous plant. *Plant Cell Tissue and Organ Culture 95:* 239-243.

Goncalves, S; Fernandes, L; & Romano, A. (2010). High-frequency in vitro propagation of the endangered species *Tuberaria major. Plant Cell Tissue and Organ Culture.* 101: 359–363.

Gonzalez, S; Guerra, PE; van Baren, C; Di Leo Lira, P; Retta, D; & Bandoni, AL. (2016). Chemical Variability of "tomillo silvestre" (*Acantholippia seriphioides*, Verbenaceae) in the Patagonian Plateau. *Boletín Latinoamericano y del Caribe de Plantas Medicinales y Aromáticas 15(1):* 61-68.

Gutterman, Y. (1994). Strategies of seed dispersal and germination in plants inhabiting deserts. *The Botanical Review 60(4):* 373-425.

Henciya, S; Seturaman, P; James, AR; Tsai, YH; Nikam, R; Wu, YC; Dahms, HU; & Chang, FR. (2017). Biopharmaceutical potentials of *Prosopis* spp. (Mimosaceae, Leguminosa). *Journal of food and drug analysis 25(1):* 187-196.

Holmgren, M; Stapp, P; Dickman, CR; Gracia, C, Graham, S; Gutiérrez, JR; Hice, C; Jaksic, F; Kelt, DA; Letnic, M; Lima, M; López, BC; Meserve, PL; Milstead, WB; Polis, GA; Previtali, MA; Richter, M; Sabaté S; & Squeo, FA. (2006). Extreme climatic events shape arid and semiarid ecosystems. *Frontiers in Ecology and the Environment 4(2):* 87-95.

İzgü, T; & Mendi, YY. (2017). The Usage of Cryopreservation and Synthetic Seeds on Preservation for Plant Genetic Resources. *International Journal of Cell Science & Molecular Biology 2(2):* 555-583.

Jackson, PW; & Kennedy, K. (2009). The global strategy for plant conservation: a challenge and opportunity for the international community. *Trends in plant science 14(11):* 578-580.

Kröpfl, AV; Deregibus, A; & Cecchi, GA. (2007). Riots in a bush steppe of the Monte: changes in vegetation. *Ecología Austral 17:* 257-268.

Larroque, AMP. (2012). *Development of a technique for micropropagation of woody species in bioreactors.* Facultad de Ciencias. Instituto Nacional de Investigación Agropecuaria. Uruguay. 83 pp. Available from: https://www.colibri.udelar.edu.uy/jspui/bitstream/123456789/3974/1/uy24-15708.pdf.

López, IS; González, FV; & Luis, JC. (2006). Micropropagation of *Helianthemum inaguae*, a rare and endangered species from the Canary Islands. *Botánica Macaronésica 26:* 55–64.

López, IS; Luis, JC; Armas, MR; & González, FV. (2004). *In vitro* propagation of *Helianthemum bystropogophyllum* Svent., a rare and endangered species from Gran Canaria (Canary Islands). *Botánica Macaronésica 25:* 71–77.

Maestre, FT; Bowker, MA; Cantón, Y; Castillo-Monroy, AP; Cortina, J; Escolar, C; Escudero, A; Lázaro, R; & Martínez, I. (2011). Ecology and functional roles of biological soil crusts in semi-arid ecosystems of Spain. *Journal of arid environments 75(12):* 1282-1291.

Mazzonia, E; & Vazquez, M. (2009). Desertification in Patagonia. *Developments in Earth Surface Processes 13:* 351-377.

McNamara, S; Tinh, DV; Erskine, PD; Lamb, D; Yates, D; & Brown, S. (2006). Rehabilitating degraded forest land in central Vietnam with mixed native species plantings. *Forest Ecology and Management 233:* 358–365.

MEA- Millennium Ecosystem Assessment. (2005). *Ecosystems and Human Well-being: Synthesis.* Island Press, Washington, DC. Available from: http://www.millenniumassessment.org.

Milesi, FA; Marone, L; Lopez de Casenave, J; Cueto, VR; & Mezquida, ET. (2002). Management guilds as indicators of environmental conditions: a case study with birds and habitat disturbances in the central Monte, Argentina. *Ecología Austral 12(2):* 149-161.

Montenegro, C; Bono, J; Parmuchi, MG; Strada, M; Manghi, E; & Gasparri, I. (2005). Deforestation and degradation of native forests. Idia XXI. *Revista de Información sobre Investigación y Desarrollo Agropecuario 5(8):* 276-279.

Mubbarakh, SA; Izhar, NA; Rajasegar, A; & Subramaniam, S. (2014). Establishment of encapsulation-dehydration technique for *in vitro* fragmented explants (IFEs) of *Rosa hybrida* L. cv. Helmut Schmidt. *Emirates Journal of Food and Agriculture 26(6):* 565-576.

Murashige, T; & Skoog, F. (1962). A revised medium for rapid growth and bioassays with tobacco tissue cultures. *Physiologia Plantarum 15:* 472-497.

Oborne, M. (2010). The bioeconomy to 2030: designing a policy agenda. OECD *Observer (278):* 35-38.

Sustainable Management of Biodiversity in Woody Ecosystems   63

Ojeda, RA; Campos, CM; Gonnet, JM; Borghi, CE; & Roig, VG. (1998). The MaB Reserve of Ñacuñan, Argentina: its role in understanding the Monte desert biome. *Journal of Arid Environments 39:* 299-313.

Peláez, DV; Bóo, RM; & Elía, OR. (1992). Emergence and seedling survival of calden in semiarid region of Argentina. *Journal of Range Management 45:* 564-568.

Peters DP; Yao J; Sala OE; & Anderson JP. (2012). Directional climate change and potential reversal of desertification in arid and semiarid ecosystems. *Global Change Biology 18(1):* 151-163.

Piñuel, L; Boeri, P; Juarez, M; Sharry, S; & Barrio, D. (2015). Bioprospecting and potential antitumor effect of protein isolates from piquillin (*Condalia microphylla*) in zebrafish (*Danio rerio). REDBIO 2015. San Miguel de Tucumán, Argentina.* Available from: http://redbioargentina.org.ar/simposio2015/resumenes/presentados/BV 67.pdf.

Quezada, F; Roca, W; Szauer, MT; Gómez, JJ; & López, R. (2005). Biotechnology for the sustainable use of biodiversity. Local capacities and potential markets. *Corporación Andina de Fomento.* 126 pp.

Raman, T; Mudappa, D; & Kapoor, V. (2009). Restoring rainforest fragments: Survival of mixed-native species seedlings under contrasting site conditions in the Western Ghats, India. *Restoration Ecology 17,* 137-147.

Rausser, GC; & Small, AA. (2000). Valuing research leads: bioprospecting and the conservation of genetic resources. *Journal of Political Economy 108(1):* 173-206.

Ravelo, A; Planchuelo, A; Abraham, E; & Navone, S. (2011). Desertification Evaluation in Argentina. *Results of the project LADA/FAO.* Gráfica Latina SA, Buenos Aires. 471 pp.

Ravetta, DA; & Soriano, A. (1998). Alternatives for the development of new industrial crops for Patagonia. *Ecología Austral 8:* 297-307.

Reynoso, MA; Sánchez Riera, A; & Vera, NR. (2016). Nutraceutical properties and safety evaluation of fruits and arrope of *Geoffroea decorticans* (Chañar). *Journal of Nutrition and Food Sciences 6,* 6:2.

Reynoso, MA; Vera, N; Aristimuño, ME; Daud, A; & Riera, AS. (2013). Antinociceptive activity of fruits extracts and "arrope" of *Geoffroea decorticans* (chañar). *Journal of ethnopharmacology 145(1):* 355-362.

Roig-Juñet, S; & Flores, GE. (2001). Geographic of the arid areas of South America Austral. Theories, concepts, methods and applications. In: Llorente-Busquets J, Morrone JJ (Eds) *Introduction to biogeography in Latin America History.* The presses of Sciences, Facultad de Ciencias. UNAM, Mexico, p 257-266.

Rundel, PW; Villagra, PE; Dillon, MO; Roig-Juñent, S; & Debandi, G. (2007). Arid and Semi-Arid Ecosystems. In: *The Physical Geography of South America.* Veblen, TT; Young, K., & Orme AE. (Eds.). Oxford University Press, 158-183 pp.

Sagar, GR; & Mortimer, AM. (1976). Approach to the study of the population dynamics of plants with special reference to weeds. *Applied Biology,* 1, 1 - 47.

Shono, K; Cadaweng, EA; & Durst, PB. (2007). Application of assisted natural regeneration to restore degraded tropical forestlands. *Restoration Ecology 15,* 620–626.

Soria, A; Esteban, J; Morales, R; Martín-Álvarez, PJ; & Sanz, J. (2008). Statistical validation of the presence in plants of chemotypes characterized by the concentration of volatile components obtained by GC-MS. Madrid. *Botánica Complutensis (32):* 225-236.

Van der Meulen, S; Brils, JM; & Borowski, I. (2012). Stakeholder interviews on ecosystem services - lessons learned with special focus on scale issues. *Presentation at the RegioResources21 Conference, Dresden, Germany,* 22.05.2012.

Verma, M; Singh, R; & Negandhi, D. (2017). Forest Ecosystem: Functions, Value and Management. In: *Ecosystem Functions and Management,* Springer, Cham. pp. 101-121.

Vijayakumar, NK; Feret, PP; & Sharik, TL. (1990). *In vitro* propagation of the endangered *Virginia roundleaf* birch (Betula uber [Ashe] Fern.) using dormant buds. *Forest Science 36(3):* 842-846.

Vilela, A; Bolkovic, ML; Carmanchahi, P; Cony, M; De Lamo, D; & Wassner, D. (2009). Past, present and potential uses of native flora and

Sustainable Management of Biodiversity in Woody Ecosystems   65

wildlife of the Monte Desert. *Journal of Arid Environments 73(2):* 238-243.

Vilela, AE; González-Paleo, L; & Ravetta, DA. (2011). Secondary metabolism of woody plants of arid zones: mechanisms of production, functions and possibilities of use. *Ecología Austral 21:* 317-327.

Villagra, P; Cony, M; Mantován, N; Rossi, B; González Loyarte, M; Villalba, R; & Marone, L. (2004). Ecology and management of algarrobos in the Monte Phytogeographyc Province. IANIGLA-IADIZA. In: *Ecology and management of native forests of Argentina.* Arturi, M., Frangi J. L., & Goya, J. F. (Eds.). Edulp, La Plata, Argentina. 32 pp.

Villagra, PE; Giordano, CV; Alvarez, JA; Cavagnaro, JB; Guevara, A; Sartor, C; Passera, CB; & Greco, S. (2011). To be a plant in the desert: water use strategies and water stress resistance in the Central Monte desert from Argentina. *Ecología Austral 21:* 29-42.

WCMC, 2001. *Protected areas program: World Heritage Sites.* UNEP World Conservation Monitoring Centre, Cambridge, United Kingdom. Available from: http://www.unep-wcmc.org/sites/wh/sundarb.html.

White, RP; & Nackoney, J. (2003). *Drylands, people, and ecosystem goods and services: a web-based geospatial analysis.* World Resources Institute. Available from: http://pdf.Wri.org/drylands pdf.

Whitford, WG. (2002). *Ecology of desert systems.* Academic Press.

In: Forest Conservation
Editor: Pedro V. Eisenlohr

ISBN: 978-1-53614-559-5
© 2019 Nova Science Publishers, Inc.

Chapter 3

FORESTS AND BRAZILIAN REPTILES: CHALLENGES FOR CONSERVATION

André Felipe Barreto-Lima[1,]* *and Melina Soledad Simoncini[2]*

[1]Departament of Physiological Sciences, Institute of Biological Sciences, University of Brasília, Brasília, DF, Brazil
[2]CICyTTP (CONICET - Prov Entre Ríos- UADER), Diamante, Entre Ríos, Argentina, Proyecto Yacaré, Laboratorio de Zoología Aplicada: Anexo Vertebrados (FHUC - UNL /MASPyMA), Santa Fe, Argentina

INTRODUCTION

Biodiversity supports life on Earth, and human beings frequently depend on biodiversity to satisfy basic needs like food, refuge, medicine, combustibles, and industrial products (Dirzo et al. 2003, Urbina-Cardona

---

[*] Corresponding Author Email: afblima1@gmail.com.

2008). However, often we come across in our studies and we question ourselves: what is the proportion of the risk that the forests and their species encounter at this critical point in their evolution? What is the speed of extinction for the threatened species that we are studying, due to effects of our actions as a global human society? Furthermore, what measures can we take to reverse the current situation, or, at least, to soften it? Finally, how can we disseminate all this information beyond the scientific community, in order to gain support from other spheres of organized civil society?

Ecologically, reptiles are essential components of the Earth's biodiversity because they play integral roles in food webs as herbivores, predators, and prey, as well as connecting aquatic and terrestrial ecosystems (Schenider et al. 2001). Nevertheless, regarding the current biodiversity crisis due various anthropogenic and environmental factors (e.g., dam constructions, hunting, deforestation, mining and climatic change), many forests and species of reptiles may disappear quickly, while others may overcome such problems, if they develop adaptive mechanisms to survive the imposed scenario on front of the global warmer and climate changes.

Considering 1,498 reptile species (The IUCN Red List assessment; Böhm et al. 2013) and to provide an overview of the risk posed to reptiles by climate change, Böhm et al. (2016) found 80.5% of species were highly sensitive to climate change, especially due to habitat specialization, while 48% had low adaptability and 58% had high exposure. Overall, 22% of species were highly vulnerable to climate change and most reptile families were found to be significantly more vulnerable to climate change than expected by chance (Böhm et al. 2016).

In addition, hotspots of climate change vulnerability did not always overlap with hotspots of threatened species richness, with most of the vulnerable species found in northwestern South America, south North America and other important world regions (Böhm et al. 2016). Therefore, those species unable to acclimate or adapt to the climatic changes will become extinct, while those that are mobile enough can survive by tracking their favored resource or habitat (Rosalino et al. 2014). In other words, the species may be flexible enough to cope to the environmental changes, explained by polymorphism of alleles, phenotypic plasticity, adaptive tolerance (physiological or behavior flexibility) or versatility that may enhance the *fitness* the animals in a new environmental condition (Potts 2004, Rosalino et al. 2014).

In this chapter, our principal objective was to address the current problems and risks for Neotropical reptile biodiversity, focusing on Squamata (lizards and snakes) and Crocodylia (caimans) from the Brazilian forest biomes, some of which are considered hotspots. We highlight the need for better conservation policy, efficient measures and methods in recent studies, in order to safeguard the herpetofauna that inhabits important forest environments, where many populations are or can be at risk since they depend fully these wild landscapes.

NEOTROPICAL HERPETOFAUNA: WHAT IS THE REAL DIMENSION OF THREAT?

Tropical forests hold more than half of the Earth's species in only 7% of the continental surface, with the Neotropical region harboring 30 to 50% of the world's herpetofauna, but little is known about the natural history and ecology of many species, making conservation strategies difficult to plan (Urbina-Cardona 2008). On the one hand, reptiles are one of the most diverse groups of terrestrial vertebrates, but on the other, they are poorly represented in biodiversity knowledge – with only 44% of reptile species evaluated (IUCN 2015). Thus, the extinction risks for Squamata (amphisbaenians, lizards and snakes) are not well understood compared to other animal groups as birds, mammals, and amphibians (Tonini et al. 2016). The global distribution of tetrapods reveals a need for targeted reptile conservation, since amphibians, birds and mammals have underpinned global and local conservation priorities, despite the distributions of reptiles representing a third of terrestrial vertebrate diversity (Roll et al. 2017).

According to the IUCN (2017), in general, the best estimate of percentage threatened species for the assessed reptiles (i.e., marine turtles, sea snakes, chameleons, crocodiles and alligators) is approximately 35%.

Currently, concerning the IUCN Red List (version 2017-3, Table 5) for the assessed species, in the Neotropical region there are 819 reptile species that are endangered (218 in Mesoamerica, 338 in Caribbean Islands, and 263 in South America, having 29 species endangered in Brazil). As of 2015, the Neotropics harbored 4,049 reptile species (1,323 in Mesoamerica, 640 in Caribbean Islands, and 2,086 in South America), representing about 39% of the 10,450 species of world's reptiles (The Reptile Database 2017). Thereby, if in the Neotropics 819 species of them are endangered, in other words, at least, 20% of all the Neotropical's reptile are at risk today!

This same estimate repeats for the absence of data on the risk of extinction of 1/5 of the reptiles of the world, listed as "Deficient Data" (DD), while one in five reptile species are threatened of extinction (Böhm et al. 2013; Bland and Böhm 2016). The assessment gaps, DD and extinction risk are greater in tropical regions – mainly in freshwater environments and on oceanic islands (Böhm et al. 2013) – and for some specific taxa and natural-histories such as fossorial (e.g., Amphisbaenia) and arboreal species (Colli et al. 2016; Tingley et al. 2016).

The phylogenetic distribution of "evolutionary distinctiveness" (ED) and threat status for Squamata has not yet reached the tipping point of extinction risk affecting most species, but immediate efforts should focus on geckos, iguanas, and chameleons, representing 67% of high-ED threatened species, and thus, the extinction risk is clustered at a broad level across the Squamata (Tonini et al. 2016).

Accordingly, lizard species in degraded tropical regions seem to be at great risk, however, the low number of threatened high-ED species in areas as the Amazon (Am) may be due to a dearth of adequate risk assessments (Tonini et al. 2016). Still, models predicted a large numbers of threatened DD species in the highly-threatened Atlantic Forest (AF) and *restinga* habitats from Brazilian cost (Böhm et al. 2013). In other words, regrettably, this means that many of the forest reptile species, of which little is known, may be threatened at this moment (e.g., Figure 1).

Figure 1. Reptile species found at Brazilian forests: a) *Iguana iguana*, b) *Enyalius brasiliensis*, c) *Leposoma baturitensis*, d) *Enyalius iheringii*, e) *Enyalius capetinga* sp nov., f) *Notomabuya frenata*, g) *Corallus hortulanus*, h) *Chironius flavolinetus*, i) *Bothrops taeniatus*, j) *Dipsas catesby*, k) *Echinanthera cephalomaculata*, l) *Atractus ronnie*, m) *Caiman latirostris*, n) *Paleosuchus palpebrosus*, o) *Paleosuchus trigonatus*, p) *Caiman crocodilus*, q) *Caiman yacare*, r) *Melanosuchus niger*. Photos: Juventino, I. (c, k, l, o, p, r); Merçon, L. (m, n, q); Migliore, S.N. (d); Pantoja, J.L. (a, f, g, j); Barreto-Lima, A.F. (e); Sifuentes, D. (i); Tourinho, P.D. (h); and Vrcibradic, D. (b).

Similarly, in Cerrado (Ce), four of 30 endemic lizard species are included in the IUCN or in Brazilian Red List of threatened species, but only one species seems to be adequately protected by the current system of protected areas (Novaes e Silva et al. 2014). For the crocodilians (*Caiman latirostris, C. crocodilus, C. yacare, Melanosuchus niger, Paleosuchus palpebrosus* and *P. trigonatus*: Figure 1 – m, n, o, p, q, r) distributed in Brazil, most have a lower risk status (Red List of IUCN 2017; http://www.iucnredlist.org/). However, the situations in each Brazilian region or biome are very different. In general, the principal threats for caimans are: habitat destruction, illegal hunting, construction of hydroelectric dams, urbanization, and pollution (Crocodile Specialist Group, Actions Plans 2010/IUCN; Campos et al. 2010; Magnusson and Campos 2010a,b; Thorbjarnarson 2010; Velasco and Ayarzagüena 2010; Verdade et al. 2010).

In resume, all these problems may occur with many species of the Brazilian herpetofauna (Figure 1), since the country has an immense territory and adds many economic, social, political and environmental problems, yielding a strong influence on any actions of preservation for the forests of important biomes such Am, AF and Ce, and its associated faunas.

BRAZILIAN FORESTS: RICHENESSES THAT ARE DISAPPEARING IN THE LANDSCAPE

Notoriously, many forests and other landscapes are disappearing or being changed rapidly, which seems a global tendency undeniable in tropical areas. The global land use and its related pressures have reduced local terrestrial biodiversity intactness, beyond its proposed planetary boundary across 58% of the world's land surface within most biomes, especially in most biodiversity hotspots and in grassland (Newbold et al. 2016). Many examples include AF, Am and Ce that are highly threatened in Brazil (see Figure 2).

Figure 2. Brazilian forests: Amazon Forest (a), gallery forest or dry forest of the Cerrado (b, c, d), and Atlantic Forest (e, f, g). Some current examples of forest impacted by anthropogenic activities: soybean plantation (h, i, j), fire (k, l) and extensive cattle raising in forest areas of Cerrado (m), and hydroelectric power plant in the Amazon region (n); Samuel Hydroelectric Power Plant, Jamari River, Candeias of Jamari municipality, Rondônia state. Photos: Barreto-Lima, A.F. (b, c); Juventino, I. (a, j, n); Pantoja, J.L. (d, h, i); Velho, D. (k, l, m); and Vrcibradic, D. (e, f, g).

In the last few decades, climate changes and global warming as consequence direct or indirect of human activities – such as fossil fuel, agricultural and livestock frontier and forest fires – have destroyed much of the world's biodiversity, mainly in Neotropical forests. Indeed, this is considered as the imminent start of 'the sixth mass extinction,' with predictions of significant additional impacts on habitats, indicating that more species will become threatened with extinction, and their distributions will move substantially, often shrinking (Thomas et al. 2004; Willis et al. 2015).

Recently, estimates reveal an exceptionally rapid loss of biodiversity over the last few centuries, indicating that a mass extinction is already under way, and averting a decay of biodiversity and the loss of ecosystem services is only possible via intensified conservation efforts, but this opportunity is rapidly closing (Ceballos et al. 2015).

Amazon Forest

The Brazilian Amazon can be characterized for having areas of dry land and seasonally floodable forests (Figure 2 a) (Haugaasen and Peres 2006), and of relevant interest for the conservation of amphibians and reptiles (Azevedo-Ramos and Galatti 2001; Vogt et al. 2001, Waldez et al. 2013). Some studies have evaluated the herpetofauna diversity in the dry land (not floodable forest; Vogt et al. 2007; Lima et al. 2008; Ávila-Pires et al. 2009; Bernarde et al. 2011), whereas few studies have described the herpetofauna of seasonally floodable forest (Gascon et al. 2000; Doan and Arizábal 2002; Neckel-Oliveira and Gordo 2004). However, the greatest diversity of species of the herpetofauna is associated to the dry land forests.

Researchers have observed how communities of herpetofauna in areas of Am present rapid responses to alterations in forest cover (e.g., Gardner et al. 2007), compared to discrete alterations such as non-selective forest cutting (Vitt et al. 1998). In the Am biome, deforestation only increased with economic and development pressures, aggravated by a worrisome reality, the rise in political power of the 'ruralists,' a coalition of landowners, soy producers, and other economic players with a powerful interest in seeing that "development" and infrastructure projects push deeper into the Am (Fearnside 2017).

In Amazonian region, year after year there are more people, roads giving them access to the forests, money pouring in for investment in agriculture and ranching, and more large projects such as hydroelectric dams (e.g., Figure 2 n): in Madeira (Jirau) and Xingu rivers (Belo Monte) there has been major deforestation – just like between the cities of Santarém and Cuiabá, where a highway was built for the soybean transport from Mato Grosso state to the ports of the Amazon River (Fearnside 2017).

Unfortunately, the advance of soybeans into former cattle pastures in Mato Grosso state, including areas that were originally savannas rather than rainforest, has been inducing ranchers to sell their land and reinvest the proceeds in buying and clearing forest areas where land is cheap, deeper in the Am region (Fearnside 2017).

Atlantic Forest

The AF is an important Neotropical biome localized in the eastern portion of the South American (Ab'Sáber 1977), mainly in Brazilian territory, and extending in Misiones, Argentina and eastern Paraguay (Galindo-Leal and Câmara 2005, Batalha-Filho and Miyaki 2011). It is a global biodiversity hotspot (Myers et al. 2000), and its original extent was 1,360,000 km^2, with 70% of the dense forest cover and the remainder by open areas (Rizzini 1997). However, less than 8% of the original AF currently exists, composed of fragments scattered along the Brazilian coast (Galindo-Leal and Câmara 2005, Batalha-Filho and Miyaki 2011). Some studies contributed on the reptile fauna from the AF, such as Sazima and Haddad (1992), Feio and Caramaschi (2002), and Marques and Sazima (2004), however it is limited to few areas in southeast Brazil (e.g., Figure 2 – e, f, g), mainly in the state of São Paulo.

It is relevant to mention that three major phylogeographic discontinuities exist in many taxon from AF, associated with glaciations and neo-tectonic activities during the Quaternary: these include bees, amphibians, reptiles, birds, bats and plants (Batalha-Filho and Miyaki 2011), and the principal hypothesis to explain the dynamic of diversification and such discontinuities is the Refuge Theory (Vanzolini and Williams 1970). These areas are dense humid forest islands isolated by open vegetation, that retracted in the periods of glacial maximum and expanded in warmer periods (Batalha-Filho and Miyaki 2011), serving as a large climate refuge for Neotropical species in the late Pleistocene, and showing high diversity of forest lizards in the center AF than southern areas (Carnaval et al. 2009). Nevertheless, it is still necessary to study more organisms in AF to understand the dynamic of diversification that generated the biodiversity of the AF (Batalha-Filho and Miyaki 2011).

The AF hotspot is one of the most threatened and fragmented biomes, and is also one of the most important for world biodiversity, due the high richness and levels of endemism (Myers et al. 2000, Galindo-Leal and Câmara 2005). Its fragmentation and threat of destruction in several regions are explained by its close location to high human demographic density areas, where the AF has reduced its vegetation drastically (Tabarelli et al. 2005), affecting the wildlife diversity in various groups of animals and plants. The impacts on the AF include human overexploitation of its resources, hunting and land exploitation, such pastures, agricultural crops, and forestry. Despite the accentuated devastation, AF is among the five regions with the highest rates of endemism of vascular plants and vertebrates (Myers et al. 2000). It also includes several endemic species of reptiles such as: chelonian *Hydromedusa maximiliani* as well as species that have been identified by anthropic occupation, such as the lizard *Liolaemus lutzae* (Machado et al. 2008), on the coast of Rio de Janeiro state, in areas of beaches where there is strong real estate speculation as threat (Franke et al. 2014). The endemism of *L. lutzae* seemed to be a consequence of a reduction of the original

Forests and Brazilian Reptiles

distribution area, and due to this high loss rate of habitat occupied, the conservation and recovery of the remaining areas affected by human actions is essential (Wink et al. 2014). Other examples are some of the species of *Enyalius* lizards with restrict distribution in the AF, such as *E. brasiliensis*, *E. iheringii* (Figure 1 b, d) and *E. perditus*, which life history is closely linked to the evolution of this biome along the eastern coast of Brazil (Barreto-Lima 2012).

Dry Forests from the Cerrado

Regarding the Brazilian Ce, it is the largest and richest savanna on Earth (Ribeiro and Walter 1998), and has a large core distribution in central Brazil with isolated patches in Am and the AF (Novaes e Silva et al. 2014). The influence of the neighboring biomes suggests patterns of geographic distribution of the species linked to these formations (Souza 2005; Uetanabaro et al. 2007). The Ce has a highly heterogeneous landscape, with vegetation patches ranging from grasslands to forests (Ribeiro and Walter 1998), a marked wet–dry seasonality, and its complexity and heterogeneity are reflected in various phytophysiognomies (Eiten 1972). In "Ce-Ca" transition, regions of floristic singularity are formed by species common to both biomes, composing complex ecosystems that harbor high diversity of native flora (Rizzini 1997).

Even considered as a kind of open *savanna*, by the predominance of grasses in a large territory, there are many Deciduous Seasonal Forests, i.e., dry forests or gallery forest (e.g., Figure 2 b, c, d,) which are important part of the mixed phytophysiognomic complex of this biome, characterized by the high density of individuals and the formation of discontinuous canopy forest formation (Ribeiro and Walter 2008). These represent potential areas of endemism, because in many areas the coverage may be of semi-deciduous forests, formations that can lead to the isolation and the speciation process of populations (Uetanabaro et al. 2007). These formations represent not only potential ecological functions for species, but are fundamental environments for the survival and maintenance of the livelihoods of many human traditional communities that use their natural resources daily (Oliveira et al. 2006). Despite the ecological-social importance of Deciduous Seasonal Forests, and their rapid conversion to agricultural landscapes (e.g., Figure 2 h, i, j), the literature on botanical characterization and ecological processes and dynamics in these environments is limited to a few research centers (Nascimento et al. 2004).

The Ce biome shelters 5% of the world's biodiversity (Mittermeier et al. 2005), but more than 50% of its area has been altered (Sano et al. 2009), even so, it is considered one of the world's *hotspots* with a serious threat of disappearing (Myers et al. 2000) due to the expansion of agribusiness, soybeans (Figure 2 j), cotton, cereals and other cash crops (Fearnside and Figueiredo 2016, Silva and Lima 2018). Also, we have observed that the environmental physiognomy of the Goiás, Mato Grosso, Mato Grosso do Sul states has been under intense environmental pressure (Figure 2 h, i, j, k, l, m), with degradation of the forested areas, mainly due to agricultural expansion in the last decades, occupying vast areas for soybean and pasture for cattle (Barreto-Lima and Simoncini *pers. obs.* 2017). Such radical environmental change is explained by the fragmentation process of forests, whose result for biodiversity is a significant reduction of the population size of species, which may lead to local extinctions (Primack 2002). Many of these forestry formations were also strongly affected by other anthropic activities, such as limestone mining, timber extraction (Uetanabaro et al. 2007), and construction of hydroelectric dams (Silvano and Segalla 2005).

In short words, most Brazilian herpetofauna is represented within the AF, Am and Ce biomes – especially in the *hotspots* (AF and Ce), since they have high level of species and endemism, exhibiting specific representatives of each biome, as well as the generalist species with wide distributions which may be found in the neighboring biomes, such as the mesic Am or dry forests of the Ca. Therefore, neglecting our forests is the same as condemning these and other species of other groups to be moved more quickly along the path of their extinctions.

OVERVIEW OF THE CURRENT IMPACTS ON THE REPTILES FROM BRAZILIAN FORESTS

In Brazil, there are 795 reptile species: 36 Testudines, 6 Crocodylia, and 753 Squamata (72 amphisbaenians, 276 "lizards" and 405 serpents) (Costa and Bérnils 2018), being the third largest country in number of reptile species in the world, just behind Australia (1,057) and Mexico (942) (Uetz and Hošek 2018). Almost half (47%) of the Brazilian reptiles are endemic to the country, mainly amphisbaenians (76%), followed by lizards (54%), serpents (40%), and chelonians (16%) (Costa and Bérnils 2018). According to the regions, the Northern is the richest in taxa of reptiles (453); Squamata (423), serpents (243), lizards (152), chelonians (25) and crocodilians (5) – this latter at equality with the Midwest, while the Northeast is the second largest for the groups, except crocodilians and snakes, and it is the lowest taxa richness in amphisbaenians (35); however, the South is the lowest richness for all groups of reptiles (Costa and Bérnils 2018).

Nevertheless, on the opposite way of biodiversity, according to the "Instituto Chico Mendes para a Conservação da Biodiversidade" (ICMBio) and the Ministry of the Environment, there are 80 species and subspecies of reptiles threatened to extinction in the official list of endangered species in Brazil – (06) chelonians, (08) amphisbaenians, (37) lizards and (29) serpents (Portaria No. 444, 2014, MMA, 2018), and up to now, no species of crocodilian. This list uses the same IUCN criteria and categories (Martins and Molina 2008), but the status of conservation is not usually the same between both lists, since the Brazilian list is based on Portaria No. 444 (MMA 2018).

Changes occur since all life exists, and the conservation biology has been mostly concerned on the anthropic actions that may affect biodiversity, as habitat destruction, population decline, and species extinction (Caro 2007). Accordingly, habitat destruction (Travis 2003) and increasingly anthropic threats arise as competition with invasive species (Yamada and Sugimura 2004). Additionally, changes in interspecific relations (Colwell et al. 2012), overexploitation (Bodmer et al. 1997), pollution (Mann et al.

2009), and oil and gas extraction have been responsible for the destruction of many pristine environments (Rosalino et al. 2014), among other many human activities.

Agriculture, Fire and Livestock

The implementation of agriculture for crop production (10,000-12,000 years ago) dramatically changed the face of the Earth (Blondel 2006), particularly after mechanization. This is one of the most impactful activities that have altered the wildlife landscapes, by cultivation of the soil for the growth of crops – agriculture –, besides cattle raising and silviculture (Rosalino et al. 2014). In addition, in many environments the vertical component of the habitat is also lost such as in soybean plantations (e.g., Figure 2 h, i, j) (Rosalino et al. 2014). The intensification of agriculture added extra challenges for species trying to live in such environments, as the increased use of many agrochemicals, most of which we do not understand all actions and impacts on the environments and wildlife, because the bioaccumulation – some of these chemicals are transported to underground and above – in ground water sheds affects mainly species that depend on these riparian environments (Rosalino et al. 2014).

The extensive livestock also is a problem that may or may not be associated to agriculture. The practice of extensive cattle raising, mainly for beef consumption, is very common in Brazil, in order to meet the demands of the domestic and foreign markets. In addition, there is another related potential problem: the indiscriminate use of fire in the forests, which is widely used by farmers (being controlled or not) for the removal of vegetation and pasture for livestock or through fires for soil cleaning and preparation of new planting (e.g., for sugar cane, corn or soybeans) (e.g., Figure 2 k, l).

As consequence, fragmentation and habitat loss are the main causes of decline and extinction of species' populations, reducing dramatically overall biodiversity (Driscoll 2004; Pardini et al. 2009). The situation is worse because there is little data on the effects of forest fragmentation and habitat loss on the reptiles, this being the least studied group (Almeida-Gomes and Rocha 2014).

Global Climate Change

Global climate change is the other major threat to biodiversity currently, due to accelerating global warming and its extremes climatic alteration (IPCC 2014), causing disturbances and deleterious impacts on global biodiversity (Pacifici et al. 2015). As most animals are ectotherms, they have body temperatures that reflect their environments to varying degrees (Angilletta et al. 2004), meaning that extremely high or low temperatures are lethal, once the temperature determines the rate of biochemical and physiological reactions; thus, *fitness* can decline by repeated exposure to deleterious temperatures (Sinclair et al. 2016). As consequences for forest reptiles, these ectotherms will need to develop morphological, physiological and behavioral responses, in order to avoid a rapid extinction, adapting to the new environmental conditions or moving to distant areas, in order to find shelters, feed, water, and other basic resources to live.

Climatic changes affect all organisms but the effect may be more important for ectotherms that have temperature-dependent sex determination (TSD), such as crocodilians (Valenzuela and Lance 2004). Thus, the increasing temperatures due to climate change IPCC (2014) may affect incubation temperature of egg-laying species, their embryonic development, sex ratios and hatching success (Piña et al. 2007, Charruau 2012), unless they modify their nesting behavior and phenology, as suggested in Simoncini et al. (2013). Eggs of *C. crocodilus* incubated at laboratory at 31°C or less produce exclusively females, at 32°C produce 95% males, although the sex ratio varies between nests (Lang and Andrews 1994). For *P. trigonatus*, eggs incubated at 31°C produced only females, and at 32°C only males, and survival is reduced at incubation temperatures less than 27°C (Magnusson et al. 1990; Campos 1993). For *C. latirostris* eggs incubated between 29 and 31°C produce 100% females, whereas eggs incubated at 33°C produce 100% males, and at 34.5°C the eggs produced females (Parachu et al. 2017); if temperatures exceed 34.5°C for a long period, embryos experience high mortality (Piña et al. 2003).

Clearly, environmental temperature affects nest temperature, and the internal nest temperature affects the proportion of females produced in wild (Simoncini et al. 2014) and the proportion of females might be affected factors such as nest location, insolation and nest plant material (Magnusson et al. 1990). Thereby, caimans and other TSD species may be susceptible to environmental changes caused by temperature, deforestation, changes of river pulse and even by the presences of pollutants, because some of them are reversal sex up to kill them, being evidences to support the biomes conservation.

Historically, crocodilian species have been exploited by humans, mainly for their skin, eggs, meat and fat (Klemens and Thorbjarnarson 1995). The characteristics of their life history make it easier to manage these reptiles as a source of protein, or cultural food for native inhabitants, or exotic food for men nowadays (Mittermeier et al. 1992). Currently, one can find six crocodilians inhabiting Brazil (Costa and Bérnils 2015): *Melanosuchus niger* (Black caiman), *Caiman crocodilus* (Spectacled caiman), *Paleosuchus palpebrosus* (Cuvier's dwarf caiman), *P. trigonatus* (Schneider's smooth-fronted caiman), *C. latirostris* (Broad snouted caiman) and *C. yacare* (Yacare caiman), all these belonging to the Alligatoridae family. Several studies refer to the absence of *M. niger* in many historical distribution areas, which could be associated with the strong hunting pressure that the species suffered (Medem 1981, 1983, Brazaitis et al. 1996 a, b).

In contrast, species of the genus *Paleosuchus* historical hunting pressure would have been low, due a low commercial value of its ossified leather adapted to terrestrial life (Medem 1981). However, both *P. palpebrosus* and *P. trigonatus* would be more affected by habitat loss because to mining activities (Campos et al. 1995), as well as deforestation, pollution, erosion, hydroelectric dams, and urbanization (Campos and Mourão 2006; Magnusson and Campos 2010). In fact, most environments inhabited by caimans are found under strong anthropogenic pressure; habitats are being

Forests and Brazilian Reptiles

modified or destroyed by activities such as urbanization, road construction, deforestation, habitat fragmentation, hydroelectric dams, pollution, and hunting (accidental or subsistence) (Magnusson 1989, Campos et al. 2013, Campos et al. 2015).

Habitat Loss and Forest Fragmentation

Regarding habitat loss due forest fragmentation, species richness can be more affected by habitat's structure and quality than the size of forest (see Almeida-Gomes and Rocha 2014 and references), e.g., in too fragmented AF, – one of the most important *hotspot* for world's biodiversity (Myers et al. 2000) – studies suggest that areas affected by habitat loss and fragmentation exhibited lizard communities that have undergone change or disappearance for the sensitive species, and increases for generalists (e.g., Urbina-Cardona et al. 2006). Thereby, in the fragments with better quality, forest and habitat-generalist species were recorded, suggesting that the greater pool of richness can result from the fragment's ability to sustain species typical of forests along with those tolerant to disturbance (Almeida-Gomes and Rocha 2014).

A wide diversity of vertebrates is threatened by losing their habitat, and due to hunting, that although forbidden in Brazil by Law 5.196/67, continues to be practiced for subsistence, and as sport or 'recreation' (Hanazaki et al. 2009, Barcellos de Souza and Nóbrega Alves 2014). Lizards such as *L. lutzae* are threatened by habitat loss (Wink et al. 2014). The official Brazilian list considers five species of Viperid snakes ('jararacas') to have some degree of threat: *Bothops alcatraz*, *B. insularis*, and *B. otavioi*, as critically endangered, which are endemic from islands of São Paulo coast (Martins and Molina 2008), and *B. pirajai* and *B. muriciensis*, as endangered (Portaria No. 444 2014).

The status list has already considered the subspecies of snakes *Lachesis muta rhombeata*, *B. bilineatus*, and *B. alternatus*, *B. cotiara*, *B. fonsecai*, *B. itapetiningae* and *B. jararacussu*, as threatened of extinction (Martins and Molina 2008). Yet, increasing urbanization is a threat, especially in eastern

Brazil, but species such as *C. latirostris* can still be found in sewage and lakes in highly urbanized areas, showing that the species is rather resistant to human impacts and that habitat modification has limited effect on the species distribution, but habitat modification determined the small size of natural populations (Moulton 1993, Freitas Filho 2007, Filogonio et al. 2010).

Concerning threat levels, *status* of conservation, and the impacts of habitat loss on Squamate biogeographic patterns in the Ce, the habitat loss and the current protected areas coverage significantly differed between biogeographic regions (Mello et al. 2015). The southernmost region is the least protected and the most impacted, with priority areas highly scattered in small, disjunctive fragments, while the northernmost region (Tocantins-Serra Geral) is the most protected and least impacted, showing extensive priority areas in all scenarios (Mello et al. 2015).

Unfortunately, the Protected Areas network is insufficient to minimize their extinction risks; only one species is fully protected and 94% of the species are under major or total conservation gaps, and the continuing rapid loss of native vegetation in the Ce hotspot indicates an urgent need for extensive conservation measures (Novaes e Silva et al. 2014). The region is under severe habitat loss due to agricultural expansion (Silvano and Segalla 2005; Silva and Lima 2018); high mechanized agriculture with a non-random deforestation accelerated by recent changes in the Brazilian National Forest Code, and thus, important regions of the Ce for the conservation of biogeographical patterns of Squamata are negatively impacted and poorly protected (Mello et al. 2015).

In the Ce biome, there are 267 Squamata species (33 amphisbaenians, 76 lizards, 158 snakes), 103 are endemics, and they represent 39% of regional richness, including 20 amphisbaenians (61% endemism), 32 lizards (42%) and 51 snakes (32%) – high endemism levels were observed within of many lineages: 47% in Tropiduridae, 57% in Gymnoph-thalmidae, 61% in Amphisbaenidae, 67% in Phyllodactylidae and Leptotyphlopidae, 80% in Elapomorphini, and 100% in Hoplocercidae (Nogueira et al. 2011). Including 10 species of Chelonia and five Crocodylia recorded (Colli et al. 2002), the overall reptilian richness of 282 species (a general endemism level of 36%) greatly surpassed values presented (Nogueira et al. 2011), and this endemism level was higher than that in other vertebrate groups (Myers et al. 2000).

The Ce is a typical plain formation with important areas of springs and swamps inhabited by caimans, which have been transformed into watering holes for cattle, while the destruction of gallery forests has caused the erosion of soil and rivers altered by anthropic activities and by intensive traffic of people and boats (Campos et al. 2015). Similarly, this situation repeats in Am and AF biomes, where commodities such as beef, soy, and sugarcane are produced through the transformation of the environment, and justify the progress of the agricultural frontiers and rural areas (Hoelle 2017).

In addition, in important transition areas or ecotonal zones under high anthropogenic pressures between Am biome and the Ce hotspot, the crocodilians are being hunted in Tocantins state (Araguaia river), and dead adult individuals of *C. crocodilus* and *M. niger* (Campos et al. 2015, Simoncini *pers. obs.* 2015) have also been found. Moreover, in Mato Grosso state (Teles Pires river) the caiman populations have been reduced to a few dispersed adult individuals, and some juveniles have been seen in auxiliary ponds of the river (Simoncini and Barreto-Lima *pers. obs.* 2017), as results of continuous, intense and growing process of human occupation and their local activities such as illegal hunting or natural habitat changes.

Hydroelectric Dams

Changes in river pulses caused by the establishment of hydroelectric dams (e.g., Figure 2 n) or the irrational use of river water for crop irrigation are other important factors that would drastically affect the nesting area availability and those areas that are not flooded after hatching of the offsprings (Thorbjarnarson 1994; Campos and Magnusson 1995; Villamarín and Suarez 2007; Villamarín et al. 2011). A recent study evaluated the effect of dam construction on the Madeira river (Amazon) in *P. palpebrosus* and *P. trigonatus*, where the authors mentioned that the dam appears to have had little effect on the use of space by the individuals that were present before the dam construction but mentioned that long term studies are necessary (Campos et al. 2017).

No less important, another problem is contamination generated by gold mining activities or agrochemical use, because heavy metals have been long recognized as a category of environmental contaminant affecting crocodilians (Schneider et al. 2012; Correia et al. 2014). These animals are top-level predators, and they may potentially accumulate toxins in their organs due to magnification; there is speculation that females have lower contamination levels due removal of these contaminants when eggs are laid (Vieira et al. 2011).

In the Am biome, the distribution of caimans reflects habitat use (Magnusson 1985; Brazaitis 1990, 1996a,b); *C. crocodilus* is distributed in floodplains, and *P. trigonatus* inhabits streams, but curiously both *M. niger* and *P. trigonatus* do not share the same environments, which could be due to mutual exclusion, although their distributions are sympatric (Rebêlo and Lugli 2001). Behaviors and physiology reduce overlap of niches between species, even between size classes of the same species (Rebêlo et al. 1997; Rebêlo and Lugli 2001; Villamarín et al. 2011). *Paleosuchus trigonatus* has highly terrestrial lifestyles (Magnusson and Lima 1991), *P. palpebrosus* colonizes large bodies of water in flooding areas (Medem 1971), while *C. crocodilus* and *M. niger* spend much of their lives in the water (Thorbjarnarson 1993, Magnusson et al. 1987).

For their growth, crocodilians eat larger preys, but they also consume small preys (Valentine et al. 1972), being generalist opportunistic predators. Thereby, caiman diets will depend on habitat quality, environmental conditions such as temperature and water levels, and the availability of prey (Da Silveira and Magnusson 1999, Richardson et al. 2002), and changes in the environment would affect their diet, affecting growth and health. In the same way, habitat does not only affect growth and health, but also affects reproduction and survival of eggs or hatchlings, as predation and flooding of nests are the main causes of egg mortality (Campos 1993).

In the central Amazonian lowlands, *P. palpebrosus* e *P. trigonatus* nest from the middle to the end of the dry season (Magnusson 1989, Magnusson and Lima 1991, Campos and Sanaiotti 2006, Campos et al. 2015) whereas *M. niger* and *C. crocodilus* nest at the height of the dry season (Campos et al. 2008; Da Silveira et al. 2010; Villamarín et al. 2011). This reproductive behavior would be associated with the flooding pulses of the Amazon River and its tributaries (Junk 1997) and so late-built nests would be susceptible to flooding (Villamarín et al. 2011). Even the degree of flooding also influences the selection of microhabitats for nesting, since the temperature of the nest will depend on its location and will determine the sexual ratio of hatchlings (Campos 1993).

Conservation of Forests and the Consequences for the Reptiles

In an era of human activities, global environmental changes, habitat loss and species extinction, conservation strategies are a crucial step towards minimizing biodiversity loss (Marchese 2015). Since global declines have long been suspected (Tingley et al. 2016), innovative means of gaining rapid insight into the status of reptiles are required to highlight urgent conservation and inform environmental policy (Böhm et al. 2013). Then, regional data gaps are apparent, being vital to understand the complete extent of distribution and extinction risk patterns of reptiles, since conservation actions can then be targeted at priority areas (Böhm et al. 2013). Basically, our knowledge on the conservation *status* of reptiles remains vastly incomplete, suggesting that many global conservation prioritization or programs have neglected reptiles (Bland and Böhm 2016, Tingley et al. 2016).

Traditionally, old decisions to create protected areas were opportunistic, based on the availability of an area for conservation, recreational values, and scenic beauty, regardless of underlying biogeo-graphical and biodiversity patterns (Pressey et al. 1993; Margules et al. 2002; Gaston et al. 2008; Mello et al. 2015). For example, in Ce biome its integral protection areas correspond to Ia Category of IUCN Protected Areas Categories System, occupying only 3% of the Ce territory (Novaes e Silva et al. 2014). The habitat loss and the current protected areas coverage significantly differed between biogeographic regions, and the patterns and processes are being erased at an accelerated pace, reinforcing the urgent need to create new reserves and to avoid the loss of the last remaining fragments of once continuous biogeographic regions (Melloet al. 2015).

The distribution predicted of threatened species reflects the distribution of recognized hotspots of biodiversity, requiring an effective conservation planning in these areas (Tonini et al. 2016). Thereby, it is essential to use the best available information and knowledge of species distributions, and when this is absent, species distribution models (Phillips et al. 2006) can be useful as a powerful tool for biological conservation.

However, we recognize it is impossible to protect all areas designated for biodiversity conservation, due to population growth, social problems, and economic and political aspects (Lips and Donelly 2005; Sarkar et al. 2006). Then, landscapes or natural environments have become semi-natural by reducing quality and losing connectivity between patches, with a decrease in size and border effects (Fahrig 2003; Fischer and Lindermayer 2007).

Therefore, to conserve herpetofauna it will be necessary to control habitat loss and increase connectivity, considering still that management and conservation areas should be carried out with social participation and scientific studies, with local production, decision-makers and policy makers (Urbina-Cardona 2008). Although, even though we may create new protected areas, we understand that it is still necessary to provide more supervision or control in many protected areas, mainly those with the largest territories and/or close to human occupied areas.

PUBLIC POLICIES, AGRIBUSINESS AND FORESTS CONSERVATION IN BRAZIL

Although the controversial revision of the Brazilian Forest Code – (BFC) was approved in 2012 by Brazilian Congress (Law of Protection of Native Vegetation, which regulates the use of the soil in private properties), only one study has so far evaluated the impacts of the new BFC on biodiversity and ecosystem services (Vieira et al. 2018). The Ce biome has undergone continuous deforestation, considering that most species of plants threatened with extinction occur in the Ce (Strassburg et al. 2017); the loss would be greater in areas in contact with the AF, to the north and southeast (Vieira et al. 2018). Furthermore, the Brazilian environmental legislation is under siege by the agribusiness lobby and interests in accelerating and weakening environmental licensing (Fearnside and Figueiredo 2016). In fact, the data are alarming and may prevent Brazil from fulfilling its environmental commitments due to a huge divergence between international agreements and internal conservation and public policy decisions (Loyola 2014; Vieira et al. 2018).

Thus, a policy mix would be necessary for a sustainable scenario such as: a) effective implementation of the BFC by governments, b) land reform, c) continuity of satellite monitoring systems, d) implementation of the Low Carbon Agriculture Plan in the Ce, e) endangered species conservation policies, f) the private sector with international certification standards, and g) a boycott of agricultural products grown in areas that are illegally deforested or of high biodiversity, such the soybean moratorium (Soares-Filho et al. 2014; Strassburg et al. 2017; Vieira et al. 2018). Inevitably, it will be necessary to decide whether to develop in the sustainable or traditional way that puts natural capital at risk (Ferreira et al. 2014; Loyola 2014). Indeed, no form of development is sustained in the long term based on highly exploitative activities, excluding people and the environment (Vieira et al. 2018).

RESEARCHES AND NEW CHALLENGES FOR CONSERVATION

In 2017, Ripple and more than 15,000 scientist signatories from 184 countries warned that the principal trouble currently is the trajectory of catastrophic climate change, due to: rising greenhouse gases (GHGs) from burning fossil fuels (Hansen et al. 2013), deforestation (Keenan et al. 2015), and agricultural production – particularly from farming ruminants for meat consumption (Ripple et al. 2014). We have unleashed a mass extinction event, the sixth in roughly 540 million years, wherein many current life forms could be extinct by the end of this century (Ripple et al. 2017). Thereby, humanity is receiving now the "second notice" that by failing to limit population growth, reassess the role of an economy rooted in growth, reduce GHGs, incentivize renewable energy, protect habitat, restore ecosystems, stop pollution and defaunation, and constrain invasive species, humanity is not taking the urgent steps needed to safeguard the biosphere (Ripple et al. 2017). These authors also warned that we urgently need changes in environmental policy, human behavior, and global inequities, but these are still far from sufficient to prevent widespread misery and catastrophic biodiversity loss; thus, humanity must practice a more environmentally sustainable alternative to business as usual (Ripple et al. 2017). From the point of view of studies with the herpetofauna, some important suggestions are addressed:

- As an important first step, it is necessary to make a uniform and stable taxonomic nomenclature – it is critical to avoid overestimation of species richness and diversity for conservation assessments, in the context of legal proceedings; thus, herpetofaunal research needs to be conducted within the appropriate socio-political and economic framework, in order to effectively implement conservation area networks (Urbina-Cardona 2008).
- New approaches to studies in biology, natural history, ecology and reptile behaviors should be employed to better recognize local and regional fauna. For example, sampling of fauna and flora in well-preserved areas and areas with agro-pastoral activity is essential for comparison and understanding of local ecologies, since such focus on surrounding areas and land use the herpetofauna is relevant to management plans, ecological zoning and monitoring, which should consider the effects of the use of such areas on the conservation unit (Uetanabaro et al. 2007).

- Mining destroyed many mountainous areas, often transforming mountain environments into plains or even lakes, polluting adjacent rivers or streams as the water emerging from the debris may contain toxic compounds (Wayland and Crosley 2006, Rosalino et al. 2014). Nevertheless, many reptile species were frequently associated with anthropogenic habitats, while others seem to depend on remnants of adjacent pristine habitats, as showed in the data on the reptiles from areas recovering from mining activity. In Upper Tocantins River, in Ce biome (Oda et al. 2017), high local species richness was observed, reflecting the increase in habitat heterogeneity due to the recovery of vegetation cover associated with the topographic re-conformation (Leite and Neves 2008), which contributed to high diversity of reptiles (Oda et al. 2017).

- Forestry practices can affect habitat characteristics that influence wildlife populations, and herpetofauna could help to understand these effects and to assist the balance of the forest management objectives and aid the forestry practices for conservation of biodiversity - i.e., studies on the reptile capture rate to predict the effects of different forestry management, and adjust them to favorable conditions for reptiles (Earl et al. 2017).

- Population and community studies at different spatial and temporal scales are necessary to understand herpetofaunal responses to anthropogenic disturbances, habitat loss and fragmentation, edge and matrix effects, and their synergy with micro-climatic gradients, emergent diseases and shifting patterns of genetic diversity (Urbina-Cardona 2008). Considering this, further research efforts are important to reach an understanding on the ecological processes driving the restoration of the original reptile diversity in recovering areas (Oda et al. 2017).

- Based on studies of climate-change scenarios, we need to reevaluate the basic role of protected area systems, in order to ensure the survival of reptile populations, identifying new actions and strategies for future conservation priorities. Thus, a monitoring plan is needed, since we understand that this is a continued and dynamic process.

- One of the important tools for addressing the protection of natural ecosystems is through the sustainable use of wildlife, where the economic benefits act as an incentive for conservation; and the identification of 'key' species of economic importance for certain ecosystems generates indirect conservation gains for other species associated to the same habitats (Larriera 2011).

Still, according to Urbina-Cardona (2008), some of the biggest challenges for herpetofaunal conservation science in the Neotropics are: a) to control habitat loss, especially in forestry areas with high diversity; b) to increase landscape connectivity along altitudinal gradients; and c) to control species invasion, as alternative species interactions can spread emergent diseases facilitated by climate change. Finally, in Brazil, many species of reptiles, as well as other animals, are overexploited, hunted and traded illegally, in addition to habitat loss for deforestation, drainage of wetlands, or intensive agriculture, and accelerated climate change. Thereby, the designation of sanctuaries or reserves could reduce the problem, but they would not be the solution.

FINAL CONSIDERATIONS

In this work, our intent was to show how and why the Brazilian forests are rapidly changing by anthropogenic activities, the main negative consequences on the forest biomes and their different groups of reptiles, the threatened species with high extinction risk, as well as those could adapt to the changes to survive in these new environments. However, it is unknown how much their natural histories could change before extinction. Indeed, we must recognize that these reptiles are still very poorly known, despite the relative volume of information available, and these animals are part of a whole system of forests (and its connection areas), which may be more or less sensitive in each one of its parts, but where the conservation effort must be focused when necessary and possible, in order to maintain the biodiversity characteristic of each biome or ecotone.

Important practices in this sense should be more encouraged and practiced in Brazil, since many tools are available for the conservation of forests, such as the declaration of protected areas, which requires a strong control, and areas conserved for their sustainable management, either for tourism or for the sustainable use of economically important species (flora or fauna). Such areas, when valorized, will directly and indirectly protect other species as well as the environments they inhabit.

It becomes clear that the Brazilian forests and their ecotonal regions need to be urgently protected in order to safeguard the species of reptiles and other groups that inhabit them, considering population monitoring actions, assessments of nearby human activities and their consequences for the forests.

ACKNOWLEDGMENTS

We appreciate the translation reviews of P. Eisenlohr and R. A. Pyron, as well as the review of the Z. Campos for improvements of the manuscript.

REFERENCES

Ab'Sáber, A. N. (1977). Os domínios morfoclimáticos da América do Sul. Primeira aproximação. [The morphoclimatic domains of South America. First approximation.] *Geomorfologia,* 52: 1-21.

Almeida-Gomes, M. and Rocha, C. F. D. (2014). Diversity and Distribution of Lizards in Fragmented Atlantic Forest Landscape in Southeastern Brazil. *J Herpetol,* 48: 423-429.

Angilletta, J. M. J., Niewiarowski, P. H., Dunham, A. E., Leaché, A. D. and Porter, W. P. (2004). Bergmann's clines in ectotherms: illustrating a life-history perspective with sceloporine lizards. *Am Nat,* 164: 168-183.

Ávila-Pires, T. C. S., Vitt, L. J., Sartorius, S. S. and Zani, P. A. (2009). Squamata (Reptilia) from four sites in southern Amazonia, with a biogeographic analysis of Amazonian lizards. *Bol Mus Para Emilio Goeldi Cienc Nat*, 4 (2): 99-118.

Azevedo-Ramos, C. and Galatti, U. (2001). Patterns of amphibian diversity in Brazilian Amazonia: conservation implications. *Biol Conserv,* 103: 103-111.

Barreto-Lima, A. F. (2012). *Distribuição, nicho potencial e ecologia morfológica do gênero Enyalius (Squamata, Leiosauridae): Testes de hipóteses para lagartos de florestas continentais brasileiras.* Doctoral thesis. Universidade Federal do Rio Grande do Sul, Porto Alegre.

Batalha-Filho, H. and Miyaki, C. Y. (2011). Phylogeography of the Atlantic Forest. *Rev Biol Biogeogr*, 31-34.

Bernarde, P. S., Machado, R. A. and Turci, L. C. B. (2011). Herpetofauna da área do Igarapé Esperança na Reserva Extrativista Riozinho da Liberdade, Acre - Brasil. *Biota Neotrop.* 11 (3): 117-144.

Bland, L. M. and Böhm, M. (2016). Overcoming data deficiency in reptiles. *Biol Conserv,* 204 (Part A): 16-22.

Blondel, J. (2006). The 'design' of Mediterranean landscapes: a millennial story of Humans and ecological systems during the historic period. *Hum Ecol,* 34: 713-729.

Bodmer, R. E., Eisenberg, J. F. and Redford, K. H. (1997). Hunting and the likelihood of extinction of Amazonian Mammals. *Conserv Biol,* 11: 460-466.

Böhm et al. (2013). The conservation status of the world's reptiles. *Biol Conserv,* 157: 372-385.

Böhm, M., Cook, D., Ma, H., Davidson, A. D., García, A., Tapley, B., Pearce-Kelly and P., Carr, J. (2016). Hot and bothered: Using trait-based approaches to assess climate change vulnerability in reptiles. *Biol Conserv,* 204 (2016): 32-41.

Brazaitis, P., Rebêlo, G. H. and Yamashita, C. (1996a). The status of *Caiman crocodilus crocodilus* and *Melanosuchus niger* populations in the Amazonian regions of Brazil. *Amphibia-Reptilia,* 17: 377-385.

Brazaitis, P., Yamashita, C. and Rebêlo, G. H. (1990). A summary report of the CITES central South American caiman study: Phase I: Brazil. In: *Crocodiles. Proceedings of the 9th Working Meeting of Crocodile Specialist Group,* IUCN - The World Conservation Union, Gland, Switzerland, 100-115.

Brazaitis, P., Rebêlo, G. H., Yamashita, C., Odierna, E. A. and Watanabe, M. E. (1996b). Threats to Pleistocene crocodilian populations. *Oryx,* 30: 275-284.

Campos, Z. (1993). Effect of Habitat on Survival of Eggs and Sex Ratio of Hatchlings of *Caiman crocodilus yacare* in the Pantanal, Brazil. *J Herpetol,* 27 (2): 127-132.

Campos, Z. and Magnusson, W. E. (1995). Relationships between rainfall, nesting habitat and fecundity of *Caiman crocodilus yacare* in the Pantanal, Brazil. *J Trop Ecol,* 11: 351-358.

Campos, Z., Coutinho, M., and Abercrombie, C. (1995). Size structure and sex ratio of dwarf caiman in the Serra Amolar, Pantanal, Brazil. *Herpetological* J, 5 (4): 321-322.

Campos, Z. and Magnusson, W. E. (2013). Thermal relations of dwarf caiman, *Paleosuchus palpebrosus*, in a hillside stream: Evidence for an unusual thermal niche among crocodilians. *J Therm Biol,* 38 (1): 20-23.

Campos, Z. and Mourão, G. (2006). Conservation status of the dwarf caiman, *Paleosuchus palpebrosus*, in the region surrounding Pantanal. *Crocodile Specialist Group Newsletter,* 25 (4): 9-10.

Campos, Z. and Sanaiotti, T. (2006). *Paleosuchus palpebrosus* (dwarf caiman) nesting. *Herpetol Rev,* (37): 81.

Campos, Z., Magnusson, W. E., Sanaiotti, T. and Coutinho, M. (2008). Reproductive trade-offs in *Caiman crocodilus crocodilus* and *Caiman crocodilus yacare*: implications for size-related management quotas. *Herp J,* 18: 91-96.

Campos, Z., Llobet, A., Piña and C. I., Magnusson, W. E. (2010). Yacare Caiman *Caiman yacare*. Pp. 23-28. In: Manolis, S. A., Stevenson, C. (eds). *Crocodiles. Status Survey and Conservation Action Plan.* Third Edition, ed. By. Crocodile Specialist Group: Darwin.

Campos, Z., Sanaiotti, T., Marques, V. and Magnusson, W. E. (2015). Geographic Variation in Clutch Size and Reproductive Season of the Dwarf Caiman, *Paleosuchus palpebrosus*, in Brazil. *J Herpetol*, 49 (1): 95-98.

Campos, Z., Muniz, F., Pires Farias, I. and Hrbek, T. (2015). Conservation status of the dwarf caiman *Paleosuchus palpebrosus* in the region of the Araguaia-Tocantins basin, Brazil. In: *Regional Reports Latin America and the Caribbean, Brazil*. Crocodile Specialist Group Newsletter, 34 (3): 6-7.

Campos, Z., Mourão, G. and Magnusson, W. E. (2017). The effect of dam construction on the movement of dwarf caimans, *Paleosuchus trigonatus* and *Paleosuchus palpebrosus*, in Brazilian Amazonia. *Plos One,* 12 (11): 1-11.

Carnaval, A. C., Hickerson, M. J., Haddad, C. F. B., Rodrigues, M. T. and Moritz, C. (2009). Stability Predicts Genetic Diversity in the Brazilian Atlantic Forest Hotspot. *Science,* 323: 785-789.

Caro, T. (2007). The Pleistocene re-wilding gambit. *Trends Ecol Evol,* 22: 281-28.

Ceballos, G., Ehrlich, P. R., Barnosky, A. D., García, A., Pringle, R. M., and Palmer, T. M. (2015). *Sci Adv,* 1-15.

Charruau, P. (2012). Microclimate of American crocodile nests in Banco Chinchorro biosphere reserve, Mexico: Effect on incubation length, embryos survival and hatchlings sex. *J Therm Biol,* 37 (1): 6-14.

Colli, G. R., Bastos, R. P. and Araújo, A. F. B. (2002). The character and dynamics of the Cerrado herpetofauna. Pp. 223-241. In: Oliveira, P. S., and Marquis, R. J. (eds). *The Cerrados of Brazil: ecology and natural history of a Neotropical savanna*. Columbia University Press, New York.

Colli, G. R., Fenker, J., Tedeschi, L. G., Barreto-Lima, A. F., Mott, T. and Ribeiro, S. L. (2016). In the depths of obscurity: Knowledge gaps and extinction risk of Brazilian worm lizards (Squamata, Amphisbaenidae). *Conserv Biol,* 204 (Part A): 51-62.

Colwell, R. K., Dunn, R. R. and Harris, N. C. (2012). Coextinction and persistence of dependent species in a changing world. *Annu Rev Ecol Evol Syst,* 43: 183-203.

Correia, J., Marsico, R., Nunes Diniz, G. T., Camargo Zorro, M. and Castilhos, Z. (2014). Mercury contamination in alligators (*Melanosuchus niger*) from Mamirauá Reservoir (Brazilian Amazon) and human health risk assessment. *Environ Sci Pollut Res,* 21 (23): 13522-13527.

Costa, H., Bérnils and R. S. (2018). Répteis do Brasil e suas Unidades Federativas: Lista de Espécies. *Herpetol Bras,* 8 (1): 11-57.

Da Silveira, R., and Magnusson, W. E. (1999). Diet of spectacled caiman and black caiman in the Anavilhanas archipelago, Central Amazonia, Brazil. *J Herpetol,* 33: 181-192.

Da Silveira, R., Ramalho, E. E., Thorbjarnarson, J. B. and Magnusson, W. E. (2010). Depredation by jaguars on caimans and importance of reptiles in the diet of jaguar. *J Herpetol,* 44: 418-424.

Dirzo, R. and Raven, P. H. (2003). Global state of biodiversity and loss. *Annu Rev Env Resour,* 28: 137-167.

Doan, T. M. and Arizábal, A. W. (2002). Microgeographic variation in species composition of the herpetofaunal communities of Tambopata Region, Peru. *Biotropica,* 34 (1): 101-117.

Driscoll, D. A. (2004). Extinction and outbreaks accompany fragmentation of a reptile community. *Ecol Appl,* 14: 220-240.

Earl, J. E., Harper, E. B., Hocking, D. J., Osbourn, M. S., Tracy, A. G., Rittenhouse, M. G. and Semlitsch, R. D. (2017). Relative importance of timber harvest and habitat for reptiles in experimental forestry plots. *Forest Ecol Manag,* 402: 21-28.

Eiten, G. (1972). The Cerrado vegetation of Brazil. *Bot Rev* 38:201-341.

Fahrig, L. (2003). Effects of habitat fragmentation on biodiversity. *Annu Rev Ecol Syst,* 34: 487-515.

Fearnside, P. M. (2017). Business as Usual: A Resurgence of Deforestation in the Brazilian Amazon. Published at the Yale School of Forestry & Environmental (https://e360.yale.edu/features/business-as-usual-a-resurgence-of-deforestation-in-the-brazilian-amazon. *Yale Environment 360*).

Fearnside, P. M. and Figueiredo, A. M. R. (2016). China's influence on deforestation in Brazilian Amazonia: a growing force in the state of Mato Grosso. Chapter 7. 229-268. In: Ray, R., Gallagher, K., López, A., Sanborn, C. (eds). *China and Sustainable Development in Latin America: The Social and Environmental Dimension*. Anthem Press. New York.

Feio, R. N. and Caramaschi, U. (2002). Contribuição ao conhecimento da herpetofauna do nordeste do estado de Minas Gerais [Contribution to the knowledge of the herpetofauna of the northeast of the state of Minas Gerais], Brasil. *Phyllomedusa,* 1 (2): 105-111.

Ferreira, J., Aragão, L. E. O. C., Barlow, J., Barreto, P., Berenguer, E., Bustamante, M., Gardner, T. A., Lees, A. C., Lima, A., Louzada, J., Pardini, R., Parry, L., Peres, C. A., Pompeu, P. S., Tabarelli, M. and Zuanon, J. (2014). Brazil's environmental leadership at risk: mining and dams threaten protected areas. *Science,* 346: 706-707.

Filogonio, R., Assis, V. B., Passos, L. F. and Coutinho, M. E. (2010). Distribution of populations of broad-snouted caiman (*Caiman latirostris*, Daudin 1802, Alligatoridae) in the São Francisco River basin, Brazil. *Braz J Biol,* 70 (4): 961-968.

Fischer, J. and Lindermayer, D. B. (2007). Landscape modification and habitat fragmentation: a synthesis. *Glob Ecol Biogeogr,* 16: 265-280.

Franke, C. R., Da Rocha, P. L. B., Klein, W. and Gomes, S. L. (2014). *Mata Atlântica e Biodiversidade. [Atlantic Forest and Biodiversity*.] Universidade Federal da Bahia. Editora da UFBA. Salvador, BA. Brasil. 461p.

Freitas-Filho, R. F. (2007). *Dieta e Avaliação de Contaminação Mercurial no Jacaré-de-Papo-Amarelo, Caiman latirostris, Daudin 1802, (Crocodylia, Alligatoridae) em dois Parques Naturais no Município do Rio de Janeiro, Brasil - Juiz de Fora. [Diet and Evaluation of Mercurial Contamination in the Alligator, Caiman latirostris, Daudin 1802, (Crocodylia, Alligatoridae) in two Natural Parks in the Municipality of*

Rio de Janeiro, Brazil - Juiz de Fora.] Master Dissertation, Universidade Federal de Juiz de Fora, Brazil.

Galindo-Leal, C. and Câmara, I. G. (2005). *The Atlantic forest of South America: biodiversity status, threats and outlook.* Island Press, Washington.

Gardner, T. A., Ribeiro-Junior, M. A., Barlow, J., Ávila-Pires, T. C. S., Hoogmoed, M. S. and Peres, C. A. (2007). The value of primary, secondary and plantation forests for a neotropical herpetofauna. *Conserv Biol*, 21: 775-787.

Gascon, C., Malcom, J. R., Patton, J. L., Silva, M. N. F., Bogart, J. P., Lougheed, S. C., Peres, C. A., Neckel, S. and Boag, P. (2000). Riverine barriers in the geographic distribution of Amazonian species. *P Natl Acad Sci*, 97 (25): 13672-13677.

Gaston, K. J., Jackson, S. F., Cantú-Salazar, L. and Cruz-Piñón, G. (2008). The ecological performance of protected areas. *Annu Rev Ecol Evol Syst*, 39: 93-113.

Hanazaki, N., Alves, R. and Begossi, A. (2009). Hunting and use of terrestrial fauna used by Caiçaras from the Atlantic Forest coast (Brazil). *J Ethnobiol Ethnomed*, 5: 36.

Hansen, J., Kharecha, P., Sato, M., Masson-Delmotte, V., Ackerman, F., Beerling, D. J., Hearty, P. J., Hoegh-Guldberg, O., Hsu, S., Parmesan, C., Rockstrom, J., Rohling, E. J., Sachs, J., Smith, P., Steffen, K., Van Susteren, L., Von Schuckmann, K. and Zachos, J. C. (2013). Assessing "dangerous climate change": Required reduction of carbon emissions to protect young people, future generations and nature. *Plos One*, 8 (art. e81648).

Haugaasen, T. and Peres, C. A. (2006). Floristic, edaphic and structural characteristics of flooded and unflooded forests in the lower Rio Purús region of central Amazonia, Brazil. *Acta Amaz*, 36 (1): 25-36.

Hoelle, J. (2017). Jungle beef: consumption, production and destruction, and the development process in the Brazilian Amazon. *J Pol Ecol*, 24: 745-762.

IPCC (2014). Climate Change 2014: Synthesis Report. Contribution of Working Groups I, II and III to the Fifth Assessment Report of the

Intergovernmental Panel on Climate Change. In: Core Writing Team, R. K., Pachauri and L. A., Meyer (eds). IPCC, Geneva, Switzerland, *The Intergovernmental Panel on Climate Change,* 151pp.

IUCN (2015). *The IUCN Red List of Threatened Species.* Version 2015.2. <http://www.iucnredlist.org>.

IUCN (2017). *IUCN Red List version 2017-3*: Table 5 Last Updated: 05 December 2017. http://www.iucnredlist.org/about/summary-statistics. Assessed: January 2018.

Junk, W. J. (1997). General aspects of floodplain ecology with special reference to Amazonian floodplains. pp. 3-20. In: Junk, W. J. (ed). *The Central Amazon Floodplain.* Berlin and Heidelberg, Germany: Ecological Studies, Springer-Verlag.

Keenan, R. J., Reams, G. A., Achard, F., Freitas, J. V., Grainger, A., and Lindquist, E. (2015). Dynamics of global forest area: Results from the FAO Global Forest Resources Assessment 2015. *For Ecol Manag,* 352: 9-20.

Klemens, M. W. and Thorbjarnarson, J. B. (1995). Reptiles as a food resource. *Biodivers Conserv,* 4 (3): 281-298.

Lang, W. and Andrews, H. V. (1994). Temperature-Dependent Sex Determination in Crocodilians. *J Exp Zool,* 270: 28-44.

Larriera, A. (2011). Ranching the broad-snouted cayman (*Caiman latirostris*) in Argentina: An economic incentive for wetland conservation by local inhabitants. pp. 86-92. In: Abensperg-Traun, M, Roe D, O'Criodain C (eds). *CITES and CBNRM. Proceedings of an international symposium on "The relevance of CBNRM to the conservation and sustainable use of CITES-listed species in exporting countries,"* Vienna, Austria, 18-20 May 2011. Gland, Switzerland: IUCN and London, UK: IIED.

Leite, F. A. S. and Neves, M. P. (2008). Reflexões sobre fechamento de mina. [Reflections on mine closure.] *E-Scientia,* 1 (1): 1-14.

Lima, A. P., Magnusson, W. E., Menin, M., Erdtmann, L. K., Rodrigues, D. J., Keller, C. and Hodl, W. (2008). *Guia de Sapos da Reserva Adolpho Ducke: Amazônia Central.* [*Adolpho Ducke Reserve Frog Guide: Central Amazonia.*] Attema Design Editorial Ltda, Manaus.

Lips, K. R. and Donnelly, M. A. (2005). Lessons from the Tropics. In: Lanoo M (ed.) *Amphibian Declines: The conservation status of United States Species*. University of California Press. USA.

Loyola, R. (2014). Brazil cannot risk its environmental leadership. *Div Dist,* 20: 1365-1367.

Machado, A. B. M., Drummond, G. M. and Paglia, A. P. (2008). *Livro Vermelho da Fauna Brasileira Ameaçada de Extinção*. [*Red Book of the Brazilian Fauna Threatened with Extinction*.] Brasília, DF: Ministério do Meio Ambiente, *2: 327-373.*

Magnusson, W. E. (1985). Habitat selection, parasites and injuries in Amazonian crocodilians. *Amazoniana,* 2: 193-204.

Magnusson, W. E. (1989). Paleosuchus. In: *Crocodiles. Their Ecology, Management and Conservation.* A special publication of the IUCN/SSC Crocodile Specialist Group. IUCN, Gland, Switzerland, 101-109.

Magnusson, W. E., Da Silva, E. V. and Lima, A. P. (1987). Diets of Amazonian crocodilians. *J Herpetol,* 2: 85-95.

Magnusson, W. E., Lima, A. P., Hero, J. M., Sanaiotti, T. M. and Yamakoshi, M. (1990). *Paleosuchus trigonatus* nests: sources of heat and embryo sex ratios. *J Herpetol,* 24: 397-400.

Magnusson, W. E. and Lima, A. P. (1991). The ecology of a cryptic predator, *Paleosuchus trigonatus*, in a tropical rainforest. *J Herpetol,* 25: 41-48.

Magnusson, W. E. and Campos, Z. (2010a). Cuvier's Smooth-fronted Caiman *Paleosuchus palpebrosus*. Pp. 40-42. In: Manolis, S. C. and Stevenson, C. (eds). *Crocodiles. Status Survey and Conservation Action Plan.* Third Edition. Crocodile Specialist Group: Darwin.

Magnusson, W. E. and Campos, Z. (2010b). Schneider's Smooth-fronted Caiman *Paleosuchus trigonatus*. Pp. 43-45 In: Manolis, S. C., Stevenson, C. (eds). *Crocodiles. Status Survey and Conservation Action Plan*. Third Edition. Crocodile Specialist Group: Darwin.

Mann, R. M., Hyne, R. V., Choung, C. B. and Wilson, S. P. (2009). Amphibians and agricultural chemicals: review of the risks in a complex environment. *Environ Pollut,* 157: 2903-2927.

Marchese, C. (2015). Biodiversity hotspots: A shortcut for a more complicated concept. *Glob Ecol Conserv,* 3: 297-309.

102 *André Felipe Barreto-Lima and Melina Soledad Simoncini*

Margules, C. R., Pressey, R. L. and Williams, P. H. (2002). Representing biodiversity: data and procedures for identifying priority areas for conservation. *J Biosci,* 27: 309-326.

Marques, O. A. V. and Sazima, I. (2004). História natural dos répteis da Estação Ecológica Juréia-Itatins. [Natural history of reptiles of Juréia-Itatins Ecological Station.] p. 257-277. In: Marques, O. A. V. and Duleba, W. (eds). *Estação Ecológica Juréia-Itatins: ambiente físico, flora e fauna.* Holos, Ribeirão Preto.

Martins, M. and Molina, F. B. (2008). Répteis. Pp. 327-373. In: Machado, A. B. M., Drummond, G. M. and Paglia, A. P. (eds). *Livro Vermelho da Fauna Brasileira Ameaçada de Extinção [Red Book of the Brazilian Fauna Threatened with Extinction]* Volume 2. Brasília, DF: Ministério do Meio Ambiente.

Medem, F. (1971). The reproduction of the dwarf caiman, *Paleosuchus palpebrosus.* 32: 159-165. In: *Crocodiles. First Working Meeting Crocodile Specialist,* IUCN Publication.

Medem, F. (1981). Los Crocodylia de Sur America Vol. 1. Los Crocodylia de Colombia. [Los Crocodylia de Colombia.] *Colciencias,* Bogota (Colombia): Ed. Carrera 7ª. Ltda.

Medem, F. (1983). *Los crocodylia de Sur America [The crocodylia of South America],* Instituto de Ciencias Naturales, Museo de Historia Natural, Universidad Nacional de Colombia, 2: 270.

Mello, P. L. H. de, Machado, R. and Nogueira, C. (2015). Conserving Biogeography: Habitat loss and vicariant patterns in endemic Squamates of the Cerrado hotspot. *Plos One,* 10 (8): 1-16.

Mittermeier, R. A., Carr, J. L., Swingland, I. R., Werner, T. B. and Mast, R. B. (1992). Conservation of amphibians and reptiles. pp. 59-80. In: Adler, K. (ed). *Herpetology: current research on the biology of amphibians and reptiles.* Oxford, Ohio: Society for the Study of Amphibians and Reptiles.

Mittermeier, R. A., Robles, P., Hoffmann, M., Pilgrim, J., Brooks, T,. Mittermeier, C. G., Lamoreux, J. and Fonseca, G. B. (2005). *Hotspots Revisited Earth's Biologically Richest and Most Endangered Ecoregions.* Conservação Internacional, 12° ed, US.

MMA - BRASIL (2018). *Biodiversidade Fauna. Ministério do Meio Ambiente*. Accessed: January 2018.

Moulton, T. P. (1993). The program of ecology of the broad-snouted caiman (*Caiman latirostris*) at CEPARNIC, Ilha do Cardoso, São Paulo, Brazil. pp. 133-134. In: Verdade, L. M., Packer, I. U., Rocha, M. B., Molina, E. B., Duarte, P. G. and Lula, L. A. B. M. (eds). *Proceedings of the 3rd Workshop on Conservation and Management of the Broad-nosed caiman* (*Caiman latirostris*), Sao Paulo, Brazil, ESALQ/USP, Piracicaba, SP, Brazil.

Myers, N., Mittermeier, R. A., Mittermeier, C. G., Fonseca, G. A. B. and Kent, J. (2000). Biodiversity hotspots for conservation priorities. *Nature,* 403: 853-858.

Nascimento, A. R. T., Felfili, J. M. and Meirelles, E. M. (2004). Florística e estrutura da comunidade arbórea de um remanescente de Floresta Estacional Decidual de encosta, Monte Alegre, GO, Brasil. *Acta Bot Bras,* 18: 659-669.

Neckel-Oliveira, S. and Gordo, M. (2004). Anfíbios, lagartos e serpentes do Parque Nacional do Jaú. pp. 161-176. In: Borges, S. H., Iwanaga, S., Durigan, C. C. and Pinheiro, M. R. (eds). *Janelas para a biodiversidade no Parque Nacional do Jaú - uma estratégia para o estudo da biodiversidade na Amazônia*.

Newbold, T., Hudson, L. N., Arnell, A. P., Contu, S., De Palma, A., Ferrier, S., Hill, S. L. L., Hoskins, A. J., Lysenko, I., Phillips, H. R. P., Burton, V. J., Chng, C. W.T., Emerson, S., Gao, D., Pask-Hale, G., Hutton, J., Jung, M., Sanchez-Ortiz, K., Simmons, B. I., Whitmee, S., Zhang, H., Scharlemann, J. P. W. and Purvis, A. (2016). Has land use pushed terrestrial biodiversity beyond the planetary boundary? A global assessment. *Science,* 353 (6296): 288-291.

Nogueira, C., Siria, R., Costa, G. C. and Colli, G. R. (2011). Vicariance and endemism in a Neotropical savanna hotspot: distribution patterns of Cerrado squamate reptiles. *J Biogeogr,* 38: 1907-1922.

Novaes e Silva, V., Pressey, R. L., Machado, R. B., Van Der Wal, J., Wiederhecker, H. C., Werneck, F. P. and Colli, G. R. (2014).

Formulating conservation targets for a gap analysis of endemic lizards in a biodiversity hotspot. *Biol Conserv,* 180: 1-10.

Oda, F. H., Ávila, R. W., Drummond, L. O., Santos, D. L., Gambale, P. G., Guerra, V., Vieira, R. S. S., Vasconcelos, T. S., Bastos, R. P. and Nomura, F. (2017). Reptile surveys reveal high species richness in areas recovering from mining activity in the Brazilian Cerrado. *Biologia,* 72 (10): 1194-1210.

Oliveira, D. A., Moreira, P. A., Melo Júnior, A. F. and Pimenta, M. A. S. (2006). Potencial da biodiversidade vegetal da Região Norte do Estado de Minas Gerais. [Potential of plant biodiversity in the Northern Region of the State of Minas Gerais.] *Unimontes Científica*, 8: 23-33.

Pacifici, M. et al. (2015). Assessing species vulnerability to climate change. *Nat Clim Chang,* 5: 215-225.

Parachu Marcó, M. V., Leiva, P., Iungman, J. L., Simoncini, M. S. and Piña, C. I. (2017). New evidence characterizing temperature-dependent sex determination in Broad-snouted caiman, *Caiman latirostris. Herpetol Conserv Biol,* 12: 78-84.

Pardini, R., Faria, G. M., Acacio, G. M., Laps, R. R., Mariano, E., Paciencia, P. A., Dixo, M. and Baumgarten, J. (2009). The challenge of maintaining Atlantic Forest biodiversity: a multi-taxa conservation assessment of an agro-forestry mosaic in southern Bahia. *Biol Conserv,* 142: 1178-1190.

Phillips, S. J., Anderson, R. P. and Schapire, R. E. (2006). Maximum entropy modeling of species geographic distributions. *Ecol Model,* 190: 231-259.

Piña, C. I., Larriera, A. and Cabrera, M. R. (2003). Effects of incubation temperature on incubation period, sex ratio, hatchling success, and survivorship in *Caiman latirostris* (Crocodylia, Alligatoridae). *J Herpetol,* 37: 199-202.

Piña, C. I., Larriera, A., Medina, M. and Webb, G. (2007). Effects of Incubation Temperature on the Size of *Caiman latirostris* (Crocodylia: Alligatoridae) at Hatching and after One Year. *J Herpetol,* 41 (2): 205-210.

Portaria No. 444, de 17 de Dezembro de 2014 (2014). (http://www.icmbio.gov.br/portal/images/stories/biodiversidade/fauna-brasileira/avaliacao-do-risco/PORTARIA_N%C2% BA_444_DE_17_DE_DEZEMBRO_DE_2014.pdf).

Potts, R. (2004). Environmental variability and its impact on adaptive evolution, with special reference to human origins. pp. 363-378. In: Rothschild, L. and Lister, A. (eds). *Evolution on Planet Earth*. Academic Press, London.

Pressey, R. L., Humphries, C. J., Margules, C. R., Vane-Wright, R. I. and Williams, P. H. (1993). Beyond opportunism - key principles for systematic reserve selection. *Trends Ecol Evol,* 8: 124-128.

Primack, R. B. (2002). Book Review: Ian Turner. 2001. The Ecology of Trees in the Tropical Rain Forest. *Garden's Bulletin Singapore* 54: 155-157.

Rebêlo, G. H., Borges, G. A. N., Yamashita, C. and Arruda Filho, A. G. (1997). Growth, sex ratio, populations structure, and hunting mortality of *Caiman yacare* in the Pantanal, Brazil. *Vida Silv Neotrop,* 6: 29-36.

Rebêlo, G. H. and Lugli, L. (2001). Distribution and abundance of four caiman species (Crocodylia: Alligatoridae) in Jaú National Park, Amazonas, Brazil. *Rev Biol Trop,* 49: 1019-1033.

Ribeiro, J. F. and Walter, B. M. T. (1998). Fitofisionomias do bioma Cerrado. [Phytophysiognomies of the Cerrado biome.] In: Sano, S. M. and Almeida, S. P. (eds). Cerrado: *Ambiente e Flora* 89-166.

Ribeiro, J. F. and Walter, B. M. T. (2008). As principais fitofisionomias do Bioma Cerrado. [The main phytophysiognomies of the Cerrado Biome.] pp. 151-212. In: Sano, S. M., Almeida, S. P., Ribeiro, J. F. (eds). *Cerrado: ecologia e flora,* Embrapa Cerrados, Planaltina, Brasil.

Richardson, K. C., Webb, G. W. J. and Manolis, S. C. (2002). *Crocodiles: Inside out: a guide to the crocodilians and their functional morph.* Surrey Beatty & Sons, Chipping Norton, NSW, Australia.

Ripple, W. J., Smith, P., Haberl, H., Montzka, S. A., McAlpine, C., and Boucher, D. H. (2014). Ruminants, climate change and climate policy. *Nat Clim Chang,* 4: 2-5.

Ripple, W. J., Wolf, C., Newsome, T. M., Galetti, M., Alamgir, M., Mahmoud, E. C., Mahmoud, I. and Laurance, W. F. 15,364 scientist signatories from 184 countries (2017). World Scientists' Warning to Humanity: A Second Notice. *Bioscience,* 67 (12): 1026-1028.

Rizzini, C. T. (1997). Tratado de Fitogeografia do Brasil: aspectos ecológicos, sociológicos e florísticos. [Treaty of Fitogeography of Brazil: ecological, sociological and floristic aspects.] *Âmbito Cultural*, 2º Ed, Rio de Janeiro, Brasil.

Roll, U., Feldman, A., Novosolov, M., Allison, A., Bauer, A. M., Bernard, R., Böhm, M., Castro-Herrera, F., Chirio, L., Collen, B., Colli, G. R., Dabool, L., Das, I., Doan, T. M., Grismer, L. L., Hoogmoed, M., Itescu, Y., Kraus, F., LeBreton, M., Lewin, A., Martins, M., Maza, E., Meirte, D., Nagy, Z. T., Nogueira, C. de C., Pauwels, O. S. G., Pincheira-Donoso, D., Powney, G. D., Sindaco, R., Tallowin, O. J. S., Torres-Carvajal, O., Trape, J. F., Vidan, E., Uetz, P., Wagner, P., Wang, Y., Orme, C. D. L., Grenyer, R. and Meiri, S. (2017). The global distribution of tetrapods reveals a need for targeted reptile conservation. *Nat Ecol Evol,* 1677-1682.

Rosalino, L. M., Verdade, L. M. and Lyra-Jorge, M. C. (2014). Adaptation and Evolution in Changing Environments. Chapter 4, 53-72. In: Verdade L. M. et al. (eds). *Applied Ecology and Human Dimensions in Biological Conservation*, Springer-Verlag Berlin Heidelberg.

Sano, E. E., Rosa, R., Silva Brito, J. L., Guimaraes Ferreira, L., Da Silva B. H. (2009). Mapeamento da cobertura vegetal natural e antrópica do bioma Cerrado por meio de imagens Landsat ETM+. [Mapping of the natural and anthropic vegetation cover of the Cerrado biome by means of Landsat ETM + images.] *Anais XIV Simposio Brasileiro de Sensoriamento Remoto, Natal, Brasil*. INPE, 1199-1206.

Sarkar, S., Pressey, R. L., Faith, D. P., Margules, C. R., Fuller, T., Stoms, D. M., Moffett, A., Wilson, K. A., Williams, K. J., Williams, P. H. and Andelman, S. (2006). Biodiversity Conservation Planning Tools: Present Status and Challenges for the Future. *Ann Rev Environ Resour,* 31: 123-159.

Sazima, I. and Haddad, C. F. B. (1992). Répteis da Serra do Japi: notas sobre história natural. [Reptiles of the Serra do Japi: notes on natural history.] p. 212-237. In: Morellato, L. P. C. (ed). *História Natural da Serra do Japi: ecologia e preservação de uma área florestal no sudeste do Brasil.* Editora da Unicamp, Campinas.

Schenider, R. L., Krasny, M. E. and Morreale, S. J. (2001). *Hands-on herpetology: Exploring ecology and conservation.* NSTA press, Arlington, Virginia. p.146.

Schneider, L., Pacheco Peleja, R., Kluczkovski, Jr. A., Martinez Freire, G., Marioni, B., Vogt, R. C. and Da Silveira, R. (2012). Mercury concentration in the Spectacled caiman and Black caiman (Alligatoridae) of the Amazon: Implications for human health. *Arch Environ Con Tox,* 63 (2): 270-279.

Silva, J. C. A. and Lima, M. (2018). Soy Moratorium in Mato Grosso: Deforestation undermines the agreement. *Land Use Policy,* 71: 540-542.

Silvano, D. L. and Segalla, M. V. (2005). Conservation of Brazilian amphibians. *Conserv Biol,* 19: 653-658.

Simoncini, M., Cruz, F. B. and Piña, C. I. (2013). Effects of environmental temperature on the onset and the extension of oviposition period of *Caiman latirostris. Herpetol Conserv Biol,* 8 (2): 409-418.

Simoncini, M., Cruz, F. B., Larriera, A. and Piña, C. I. (2014). Effects of climatic conditions on sex ratios in nests of Broad-snouted caiman. *J Zool,* 293: 243-251.

Sinclair, B. J., Marshall, K. E., Sewell, M. A., Levesque, D. L., Willett, C. S., Slotsbo, S., Dong, Y., Harley, C. D. G., Marshall, D. J., Helmuth, B. S. and Huey, R. B. (2016). Can we predict ectotherm responses to climate change using thermal performance curves and body temperatures? *Ecol Letters*, 1-14.

Soares-filho, B., Rajão, R., Macedo, M., Carneiro, A., Costa, W., Coe, M., Rodrigues, H. and Alencar, A. (2014). Cracking Brazil's Forest Code. *Science,* 344: 363-364.

Souza, F. L. (2005). Geographical distribution patterns of South American side-necked turtles (Chelidae), with emphasis on Brazilian species. *Rev Esp Herp,* 19: 33-46.

Souza, J. B. and Nóbrega Alves, R. R. (2014). Hunting and wildlife use in an AF remnant of northeastern Brazil. In: Franke, C. R., Da Rocha, P. L. B., Klein, W., Gomes, S. L. (eds). Mata Atlântica e Biodiversidade. [Atlantic Forest and Biodiversity.] *Trop Conserv Sci,* 7 (1): 145-160.

Strassburg, B. B. N., Brooks, T., Feltran-Barbieri, R., Iribarrem, A., Crouzeilles, R., Loyola, R., Latawiec, A. E., Oliveira Filho, F. J. B., Scaramuzza, C. A. de M., Scarano, F. R., Soares-Filho, B. and Balmford, A. (2017). Moment of truth for the Cerrado hotspot. *Nat Ecol Evol,* 1:1-3.

Tabarelli, M., Pinto, L. P., Silva, J. M. C., Hirota, M. and Bede, L. (2005). Challenges and opportunities for biodiversity conservation in the Brazilian Atlantic Forest. Conserv Biol, 19: 695-700.

Thomas, C. D., Cameron, A., Green, R. E., Bakkenes, M., Beaumont, L. J., Collingham, Y. C., Erasmus, B. F. N., Ferreira de Siqueira, M., Grainger, A., Hannah, L., Hughes, L., Huntley, B., van Jaarsveld, A. S., Midgley, G. F., Miles, L., Ortega-Huerta, M. A., Townsend, P. A., Oliver, L., Phillips, O. L. and Williams, S. E. (2004). Extinction risk from climate change. *Nature,* 427: 145-148.

Thorbjarnarson, J. B. (1993). Diet of the spectacled caiman (*Caiman crocodilus*) in the Central Venezuelan Llanos. *Herpetologica,* 49: 108-117.

Thorbjarnarson, J. B. (1994). Reproductive ecology of the spectacled caiman (*Caiman crocodilus*) in the Venezuelan Llanos. *Copeia,* 907-919.

Thorbjarnarson, J. B. (2010). Black Caiman *Melanosuchus niger.* Pp. 29-39. In: Manolis, S. C. and Stevenson, C. (eds). *Crocodiles. Status Survey and Conservation Action Plan.* Third Edition, Crocodile Specialist Group: Darwin.

Tingley, R., Meiri, S. and Chapple, D. G. (2016). Addressing knowledge gaps in reptile conservation. *Biol Conserv,* 204 (Part A): 1-5.

Tonini, J. F. R., Beard, K. H., Ferreira, R. B. and Jetz, W., Pyron, R. A. (2016). Fully-sampled phylogenies of squamates reveal evolutionary patterns in threat status. *Biol Conserv*, 204: 23-31.

Travis, J. M. J. (2003). Climate change and habitat destruction: a deadly anthropogenic cocktail. *Proc Roy Soc Lond B Biol Sci,* 270: 467-473.

Uetanabaro, M., Souza, F., Landgref, P., Figueira, A., Albuquerque and R. (2007). Anfíbios e répteis do Parque Nacional da Serra da Bodoquena, Mato Grosso do Sul, Brasil. [Amphibians and reptiles of the Serra da Bodoquena National Park, Mato Grosso do Sul, Brazil.] *Biota Neotropica,* 7 (3): 279-289.

Uetz, P. and Hošek, J. (2015). *The Reptile Database*. Available at: http://www.reptile-database.org/.

Urbina-Cardona, J. N. (2008). Conservation of Neotropical Herpetofauna: Research Trends and Challenges. *Trop Conserv Sci,* 1 (4): 359-375.

Urbina-Cardona, J. N., Olivares-Pérez, M. and Reynoso, V. H. (2006). Herpetofauna diversity and microenvironment correlates across the pasture-edge-interior gradient in tropical rainforest fragments in the region of Los Tuxtlas, Veracruz. *Biol Conserv,* 132: 61-75.

Valentine, J., Walther, Jr. J. and Larry, M. I. (1972). Alligator diets on the Sabine National Wildlife Refuge, Lousiana. *J Wildlife Mange,* 36 (3): 809-815.

Valenzuela, N. and Lance, V. (2004). *Temperature-dependent sex determination in vertebrates.* Smithsonian Books. Washington, USA.

Vanzolini, P. E. and Williams, E. E. (1970). South American anoles: the geographic differentiation and evolution of the *Anolis chrysolepis* species group (Sauria: Iguanidae). *Arq Zool,* 19: 1-298.

Velasco, A. and Ayarzagüena, J. (2010). Spectacled caiman *Caiman crocodilus* Pp. 10-15 In: Manolis, S. C. and Stevenson, C. (eds). *Crocodiles. Status Survey and Conservation Action Plan.* Third Edition. Crocodile Specialist Group: Darwin.

Verdade, L. M., Larriera, A. and Piña, C. I. (2010). Broad-snouted Caiman *Caiman latirostris.* Pp. 18-22 In: Manolis, S. C. and Stevenson, C. (eds). *Crocodiles. Status Survey and Conservation Action Plan.* Third Edition. Crocodile Specialist Group: Darwin.

Vieira, L. M., Nunes, V. da S., Amaral, M. C. do A., Oliveira, A. C., Hauser-Davis, R. A. and Campos, R. C. (2011). Mercury and methyl mercury ratios in caimans (*Caiman crocodilus yacare*) *J Environ Monit,* 13: 280-287.

Vieira, R. R. S., Ribeiro, B. R., Resende, F. M., Brum, F. T., Machado, N., Sales, L. P., Macedo, L., Soares-Filho, B. and Loyola, R. (2018). Compliance to Brazil's Forest Code will not protect biodiversity and ecosystem services. *Div Dist.,* 24: 434-438.

Villamarín-Jurado, F. and Suárez, E. (2007). Nesting of the Black Caiman (*Melanosuchus niger*) in Northeastern Ecuador. *J Herpetol,* 41: 164-167.

Villamarín, F., Marioni, B., Botero-Arias, R., Thorbjarnarson, J. B. and Magnusson, W. E. (2011). Conservation and management implications of nest-site selection of the sympatric crocodilians *Melanosuchus niger* and *Caiman crocodilus* in Central Amazonia, Brazil. *Biol Conserv,* 144: 913-919.

Vitt, L. J., Ávila-Pires, T. C. S., Caldwell, J. P. and Oliveira, V. R. L. (1998). The impact of individual tree harvesting on thermal environments of lizards in Amazonian rain forest. *Conserv Biol,* 12: 654-664.

Vogt, R. C., Moreira, G. M. and Oliveira, C. D. (2001). Biodiversidade de répteis do bioma floresta amazônica e ações prioritárias para sua conservação. [Biodiversity of reptiles of the Amazon forest biome and priority actions for its conservation.] p. 89-96. In: Capabianco, J. P. O. R. (ed). *Biodiversidade na Amazônia Brasileira Estação Liberdade*, Instituto Socioambiental, São Paulo.

Vogt, R. C., Ferrara, C. R., Bernhard, R., Carvalho, V. T., Balensiefer, D. C., Bonora, L. and Novelle, S. M. H. (2007). Herpetofauna. Pp. 127-143. In: Py-Daniel, L. R., Deus, C. P., Henriques, A. L., Pimpão, D. M. and Ribeiro, O. M. (eds). *Biodiversidade do Médio Madeira: Bases científicas para propostas de conservação*. INPA, Manaus.

Waldez, F., Menin, M. and Vogt, R. C. (2013). Diversidade de anfíbios e répteis Squamata na Região do Baixo Rio Purús, Amazônia Central, Brasil. [Diversity of Squamata amphibians and reptiles in the Lower Purús River Region, Central Amazonia, Brazil.] *Biota Neotrop,* 13 (1): 300-316.

Wayland, M. and Crosley, R. (2006). Selenium and other trace elements in aquatic insects in coal mine–affected streams in the Rocky Mountains of Alberta, Canada. *Arch Environ Contam Toxicol,* 50: 511-522.

Willis, S. G., Foden, W., Baker, D. J., Belle, E., Burgess, N. D., Carr, J. A., Doswald, N., Garcia, R. A., Hartley, A., Hof, C., Newbold, R., Smith, R. J., Visconti, P., Young, B. E. and Butchart, S. H. M. (2015). Integrating climate change vulnerability assessments from species distribution models and trait-based approaches. *Biol Conserv*, 190: 167-178.

Wink, G. R., Almeida-Santos, P. and Rocha, C. F. D. (2014). Potential distribution of the endangered endemic lizard *Liolaemus lutzae* Mertens, 1938 (Liolaemidae): are there other suitable areas for a geographically restricted species? *Braz J Biol*, 74: (2) 338-348.

Yamada, F. and Sugimura, K. (2004). Negative impact of an invasive small Indian mongoose *Herpestes javanicus* on native wildlife species and evaluation of a control project in Amami-Ohshima and Okinawa Islands, Japan. *Glob Environ Res*, 8: 117-124

In: Forest Conservation
Editor: Pedro V. Eisenlohr

ISBN: 978-1-53614-559-5
© 2019 Nova Science Publishers, Inc.

Chapter 4

PRIORITIZATION OF AREAS FOR PERMANENT PRESERVATION FOR FOREST RECOVERY AIMING LANDSCAPE CONNECTIVITY

Emanuelle Brugnara[1,], Vinícius de Freitas Silgueiro[1] and Julio Cesar Wojciechowski[2]*

[1]Instituto Centro de Vida (ICV),
Alta Floresta, MT, Brazil
[2]State University of Mato Grosso,
Faculty of Agrarian and Biological Sciences,
Alta Floresta, MT, Brazil

INTRODUCTION

There are several modes for land use and land occupation, which have been gaining space in different scenarios of landscape, replacing plant cover and endangering biodiversity. This tendency of landscape transformation is

---

[*] Corresponding Author Email: emanuelle.brugnara@gmail.com.

fundamental to establish protected areas and restore degraded areas with the aim to protect biodiversity in fragmented landscapes under human intervention, where the success of biodiversity conservation depends on the capability of biota survival in fragmented landscapes under man intervention (Bennett 2003).

With the increase of agriculture expansion, there was a need to create more effective and restrictive laws aiming to establish actions for a sustainable production and preservation of natural resources. In order to insert a property inside legal requirements, it is necessary to map all the area, making possible to quantify irregular and occupied or degraded areas, reminiscent of native vegetation and Areas for Permanent Preservation (APP) so that it is posteriorly implemented an action plan for recovery.

Due to the consequences of agricultural expansion, forest fragmentation stands out as a major threat to biodiversity conservation, with large-scale effects resulting in heterogeneous landscapes composed of isolated remnants (Jesus 2013). The forest fragmentation involves transformation of a large habitat into smaller places than the original area (Wilcove et al. 1986), which may lead to the isolation of the remaining natural formations and populations, a modified gene flow and structure, intensification of competition and extinction of species (Metzger 1999).

The constant search for solutions aiming to diagnose and management of biodiversity, given the relative lack of data connected to these processes, matches with the development of methods and tools even more efficient in Geographical Information Systems (GIS), with application on the solution of the environmental problems (Paese et al. 2012). In addition, the survey of satellite images has become essential for monitoring environmental parameters, aiming to obtain the necessary data for the plan and management of the territory (Horning et al. 2010; Pettorelli et al. 2014; Buchanan et al. 2009). Furthermore, the use of GIS combined with spatial analysis and tools possibly create strategies that combine socioeconomic development and conservation of natural resources, involving a complexity of processes and priorities. In relation to forest recovery of degraded APP, they can contribute to the short, medium and long-term plans aiming the success of recovery actions.

Forest restoration aims to increase the potential for self-recovery of certain degraded areas, through plantations with high diversity of native species from the region. Initiatives are generally applied for restoration with the purpose of obeying the environmental legislation, reestablish the ecosystem services and local conservation of biodiversity. Considering these purposes, there is the addition of the necessity and opportunity to potentially increasing the available habitat in anthropized landscapes. In this context, biological diversity is introduced as an important target in restoration actions.

Forest recovery can be considered as one of the great challenges for the implementation of Law No. 12,651 of May 25, 2012, known as the "New Brazilian Forest Code." In view of the variables mentioned by the New Code, which define the quantity of forest to be recomposed in each situation, new discussions appear and are often oriented on the technologies and tools available for planning and monitoring areas under forest restoration (Silgueiro et al. 2017).

The difficulties encountered in restoration actions, mainly in Areas for Permanent Preservation (APP), are related to the complexity of the implementation and monitoring of processes. For the delimitation of Areas for Permanent Preservation (APP) to be restored it is necessary to obtain the dimensions of degraded natural resources, the size of the property and date when the area had its native vegetation suppressed. According to Tambosi et al. (2016), due the great territorial extension that comprises the municipalities from the state of Mato Grosso, Brazil, it is of great importance the use of automated tools for diagnosing the conservation status of the APP, with the purpose of guiding the decision-making directed to the actions for restoration of degraded areas.

In the state of Mato Grosso, actions for forest recovery are conducted under the initiatives of the society and local public power. Different techniques applied for the implementation of the recovery and monitoring of forest are often tested by the Instituto Centro de Vida (ICV), a Civil Society Organizations in the Public Interest (OSCIP), which develops the diagnostics that comprise economic, cultural, social and environmental issues for the reinforcement of sustainable productive chains with a partnership (Hoffmann 2015; Silgueiro et al. 2015; ICV 2010).

The increased connectivity of the landscape can be used as an important tool for decision making on where should start restoration actions (Tambosi 2014), allowing the connection between fragments to be reestablished and, therefore, occurs the return of flows between their elements, increasing the chance of survival for biological communities and their species. According to Tambosi et al. (2016), the connectivity in the landscape causes a greater environmental gain directly related to a lower cost still in the initial phases of restoration actions. This approach is directed to the analysis of landscape structure and it is of great value, mainly in places where there are a considerable number of areas to be restored, and it is necessary to choose priorities.

The prioritization of the Delimited Area for Permanent Preservation to be restored can be characterized such as an intervention strategy, since it allows identifying areas, which contribute to the connectivity and biological flow, setting them as potential ecological corridors. Priority should be given to techniques that ensure conservation of areas with greater fragility, stability and maintenance of functionalities of each environment, as well as the increase in the connectivity with the purpose of minimizing the effects of ecosystem fragmentation (Muchailh et al. 2009).

Areas already restored in regions with a higher availability of habitats and with a greater potential for flow of organisms are more likely to be re-colonized by individuals from the fauna, increasing the arrival of seeds and, therefore, the chances of recovery success (Crouzeilles and Curran 2016). Thus, it is of great importance the analysis of landscape models and metrics that express the connectivity degree and biological flow in forest areas in process of restoration.

Landscape metrics refer to algorithms that quantify applied peculiarities at the isolated fragment level, fragment classes or across the landscape extent. Landscape analysis by using metrics on forest fragments represents a considerable advance in the interpretation of spatial relationships (Rezende 2011). These metrics allow the criterious measurement of the landscape, providing relevant parameters for the consolidation and integration in the system of free spaces (Bunn et al. 2000).

The results obtained in analyzes of landscape metrics contribute to the plan and management of the environment of the land as a whole, providing new information for forest restoration and integration of fragmented areas. An approach that has achieved space in analysis of connectivity is the use of graph theory in order to calculate the measures of habitat availability as indicators of landscape connectivity (Bunn et al. 2000; Urban and Keet 2001).

An important factor for restoration actions is the knowledge of conditions and history of the area to be recovered, in order to predict the dynamics of the area and choice of the most appropriate intervention methods. In addition, practical approaches identifying and quantifying appropriate sites to maintain or re-establish connectivity are necessary to plan achievable goals and comply the environmental legislation (Wimberly 2006).

In rural settlement projects, despite the efforts in the legal field, there are still problems related to the environmental management. The majority of the projects do not present a plan and the choice of the areas is performed in an inadequate way, without observing the characteristics of the place (Nascimento Soares 2008). Thus, despite social advances that the settlements represent and their relevance into minimizing their role in deforestation, due to the lack of plans and effective environmental legislation, settlers hardly respect the Areas for Permanent Preservation or legal reserves.

The overview for environmental regularization in settlements projects has evolved slowly, since there is still a lack of clarity on the part of public power in relation to the stages and coordination of the registrations and validations of the Environmental Rural Register (CAR, from Portuguese jargon). This fact also results in the difficulty of recomposing environmental liabilities, preventing the socioeconomic advance of family agriculture in the state. Thus, the public remains with a consequent liability and environmental embargoes, where there is a legal impediment of any form of land use and soil occupation, making impossible the access and agricultural production (ICV 2016).

Future expectations are related to data validation generated by Environmental Rural Register, since they serve as a tool to perform analyses and studies aimed at new perspectives in relation to the compliance with legislation and its improvements, opening possibilities for the application of conservation measures of landscapes at effective scales.

Considering the relevance of studies with the purpose of analyzing the landscape, studies including the plan for forest restoration of degraded areas are of great value. In this context, the purpose of this work was to identify the Area for Permanent Preservation of watercourses and degraded headwaters and to consider the connectivity of landscape and biological flow to prioritize forest restoration of these areas in the Settlement Project São Pedro, Brazil.

MATERIAL AND METHODS

Study Area

The area of study corresponds to the Settlement Project São Pedro (Figure 1), localized in the geographical coordinates 56° 41' 22,212" West and 9° 46' 10,474" South, in the North region of the State of Mato Grosso, in the municipality of Paranaíta, with a total area corresponding to 34,894.84 hectares.

The Settlement Project São Pedro was funded in the year 1997 by The National Institute of Colonization and Agrarian Reform (INCRA) with the capacity to settle approximately 776 families. The economy is sustained by livestock farming, mainly milk, and its territory is divided into 22 rural communities, with a variable number of plots (Oliveira and Bergamasco 2014), with average sizes around 40 hectares.

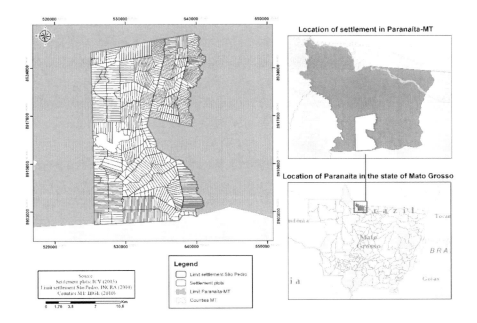

Figure 1. Localization map of the Settlement Project São Pedro.

In the region of the study area, the predominating forest phytophysionomy is an Open Ombrophilous Forest, characterized by a vegetation transition between Amazon Forest and areas beyond Amazon forests. There is three physiognomies dominated with typical genus, among them Palm Trees represented by *Attalea speciosa* Mart. Ex Spreng, which is a Babassu Palm, and the *Attalea maripa* (Aubl.) Mart. (inajá palm), bamboo forest comprised of individuals of *Guadua superba* Huber. (Small bamboo) and the sorococa forest with individuals of *Phenakospermum guianensis* (A. Rich.) Endl. Ex Miq. (Sororoca forest) (IBGE 2012).

The climate is equatorial, classified as *Am* with an average precipitation of 2.500 mm/year according to the Köppen climate classification (Bernasconi et al. 2009). In the region, there is Dystrofic Red-Yellow Argisol, occurring in low quantities of Red-Yellow Latosol and Yellow Latosol (SIBCS 2013).

Databases Accessed

For generating candidate areas for recovery, which are known as Delimited Areas for Permanent Preservation or DAPP, they were initially reunited to all features associated to hydric resources, soil covering and the lots limits that comprise the Settlement São Pedro. This candidate areas for recovery were elaborated, adapted and organized by the Geotechnologies Nucleus from Instituto Centro de Vida (ICV), based on the necessary classes for elaborating the Environmental Rural Register. They have a scale of 1:25,000 and the reference year 2016.

The delimitation of candidate areas for restoration was performed in the software ArcGIS 10.1 and mapped with auxiliary images from Pleiades 1A satellite with a spatial resolution of 50 centimeters, from Landsat-5 and Landsat-8 considering a spatial resolution of 30 meters and from the satellite Spot-5 sensor HRG with spatial resolution of 2.5 meters.

The basis of hydrographic basins comprehends the following components: headwaters, lakes or natural lakes until 20 hectares, a shapefile containing the total water surface from the settlement, water course of simple margin (with width until 10 meters) and water courses with 10 to 50 meters.

For determining the classes of soil cover, it was used a supervised classifier by maximum likelihood (MAXVER) when using the tool "Majority filter" by collecting samples based on the visual interpretation and on classes of the Environmental Rural Register. Representative classes were obtained from remaining native vegetation, in areas of consolidated use that were converted to the alternative use of soil until July 22 in 2008, also deforested areas after July 22 in 2008 and rocky outcrops.

Besides the variables related to the hydrography and soil cover, it was considered the size of the fiscal module used in the municipality where the study was conducted for delineating rural properties. The fiscal module involves a measure of the agricultural unity in hectares, with the aim of establishing a minimum standard for demonstrating its viability as a productive unity, depending on its localization (Brasil 1964).

Pursuant to Law 6,746/79 and Law 8,629/93, it is necessary to take into account the type of the predominant exploitation in the municipality and its financial income in order to determine the size of a fiscal module, as well as other incomes considered expressive in relation to the income or area used (Incra 1993).

At the study site, the size of the fiscal module is 100 hectares. The properties were separated in areas considering the following ranges of fiscal modules: up to 1 fiscal module, from 1 to 2 fiscal modules, from 2 to 4 fiscal modules, from 4 to 10 fiscal modules, and above 10 fiscal modules. Since the study area is a settlement comprised of lots, the size of the fiscal module is applied to each lot individually.

Delimitation of Candidate Areas for Recovery

From all the features of hydrography, soil cover and boundaries of the property, Areas for Permanent Preservation (APP) and Delimited Areas for Permanent Preservation (DAPP) were generated in an automated way, with the Delimited Areas being the candidate areas for restoration. For this purpose, a structured model in the software ArcGIS 10.1 was used, in its Modelbuilder function, which was created by the ICV Nucleus of Geotechnology and elaborated according to the parameters established by the Forest Code.

In the Modelbuilder used (Figure 2), basically the following five ArcToolbox extension tools are adopted in ArcGIS 10.1: buffer, union, clip, erase and dissolve. According to Butturi et al. (2017), this model was developed from flows that link a sequence of tools and databases, allowing the elaboration of work routines, as well as the automated processing of territorial extensions, facilitating work in large areas such as municipalities and including states. After the generation of the Areas for Permanent Preservation and of the Delimited Areas for Permanent Preservation, a review of the data collected was performed, aiming to remove possible inconsistencies generated during the processing, which increases the quality of products.

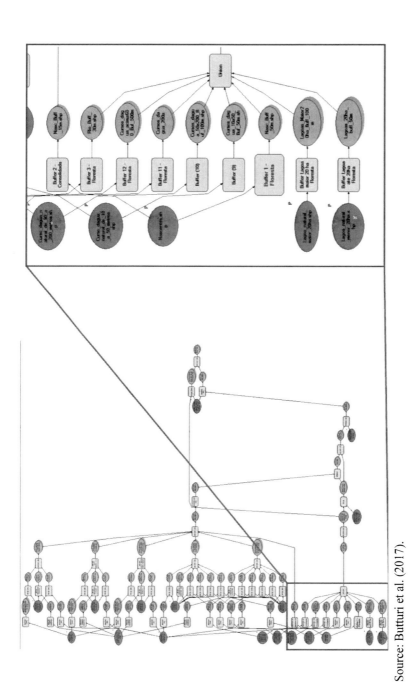

Source: Butturi et al. (2017).

Figure 2. Processing steps organized in the software "*Modelbuilder*" for automatic generation of the Areas for Permanent Preservation (APP) and of the Delimited Areas for Permanent Preservation (DAPP).

In order to avoid the overlap between polygons, the "dissolve" tool was used in the software ArcGIS 10.1, in order to merge adjacent polygons of the same class. The file generated was organized in the attributes table to contain a column with the unique identifier of polygons (ID), one with the calculated area of each polygon and another one indicating if the polygon was a fragment already existing in the landscape (with value 1) or a candidate area for restoration (with value 0).

Graph Theory and Integral Index of Connectivity

The evaluation of the contribution of each fragment in the maintenance of connectivity was carried out with the use of the approach on graph theory to calculate the measures for habitat availability as landscape indicators of connectivity in the software Conefor Sensinode 2.6.

The graph theory allows each fragment to be represented as nodes. The fragments that are linked together among them represent functional connectivity, using as a criterion the dispersion capacity of the focal organism. Therefore, the fragments of the landscape comprise a network of elements interacting among them, which integrate the information of each fragment, allowing the identification of functionally connected areas and key fragments for the maintenance of the landscape connection (Proulx et al. 2005).

In studies based on graph theory, a landscape may show different graph structures depending on the species studied and their dispersion capacity (Paese 2002). Therefore, species with low dispersion capacity result in a smaller number of connections between the nodes and a higher number of components or sub-graphs when related to the same landscape represented as graph for species with a greater dispersion (Tambosi 2014).

It is evidenced the importance of performing analyzes focusing on different capacities for organisms dispersion presented in the study area, in order to verify the capacity of species displacement, as well as, the limit of distance in which individuals are able to reach.

In order to perform the analyzes proposed in this study, an index describing the graph theory known as Integral Index of Connectivity (IIC), and its IICflux fractions that are related to the flux from the environment and of genes, and IICconnector, linked to the relevance of the node as a unique connection between fragments, proposed by Saura and Rubio (2010).

The IIC is an index based on the binary model that allows to verify the presence or absence of connections with neighboring fragments and the importance of inserting a fragment into the landscape through restoration measures. The IIC value can range from 0 to 1, increasing as the landscape becomes more connected. The calculus is performed by summing the products of the attributes of a given pair of connected nodes, divided by the number of connections of the shortest path between those two nodes. In this context, each node presents a contribution to the IIC value of the graph (Tambosi 2014).

The calculus of the IICflux and IICconnector fractions and their respective variations (varIICflux and varIICconnector) allows comprehending the potential flow of organisms between fragments and their importance for the maintenance or increase of landscape connectivity. In situations where the area to be restored creates a new connection between isolated fragments containing high values of varIICconnector, it was considered to be a priority area for connection. Some areas were determined to be important for biological flow, since they are connected to an already existing fragment, with a potential biological flow and colonization with high values of varIICflux. The high values for both variables were considered to be of the highest priority (Silgueiro et al. 2017).

Analysis of Fragments Containing Native Vegetation and Calculus of the Euclidian Distance

In order to verify the minimum distance for connection of fragments in the area of study, the landscape was previously analyzed by using the tool of the Euclidean Distance available in the software ArcGIS 10.1. The calculus of the Euclidean Distance was performed by using the shapefile containing fragments of native vegetation with a pixel size corresponding to 10 meters. The classification method by the Euclidean distance is a supervised procedure, on which the Euclidean distance is used to associate a pixel with a given class (Cruz and Ribeiro 2008).

From the result obtained in this analysis, a visual analysis was carried out for fragments of native vegetation with the purpose of verifying the maximum distances between fragments found in the Settlement Project São Pedro, inferring about possible implications for landscape connectivity and biological flow and, therefore, for forest recovery in the Delimited Areas for Permanent Preservation (DAPP).

Input Files Generation in Conefor with the File Extension Conefor Inputs

The input files for processing candidate areas for restoration in the software Conefor Sensinode 2.6 software were generated using the extension "Conefor Inputs" in the software ArcGIS 10.1. It was selected the previously generated shapefile which comprise fragments with native vegetation and the Delimited Area for Permanent Preservation. The table of attributes contains a column with the ID of the fragments and the table with the areas corresponding to each fragment.

In this analysis, the calculus of the distances was restricted to a limit distance of 500 meters, selected according to the results obtained in the calculus of the Euclidean distance. It was opted to calculate the Euclidean distance from edge fragments, in order to subsequently analyzing the ability of an organism to cross an area with no habitat during the movement to another fragment.

Based on the options selected, the extension "Conefor Inputs" generated two text files. One of these text files is called "nodes," having columns with the ID and fragment attributes, and the other with the prefix "distances," encompassing the distance between fragments pairs of the landscape.

It was necessary to generate a new "nodes" file containing three columns, in order to obtain the opportunity to make simulations between the fragments. The first column from the file have the ID of the fragments, the second column have the area of fragments and the third column indicates if the fragment was already in the landscape (with value 1) or if they would be created in the landscape (with value 0). For the generation of this new file, the attribute table from the shapefile file was exported into an input file from the software Conefor and it was subsequently edited to have only the three columns needed.

Simulation for Different Dispersion Abilities

Due to the great variety of species able to move at different distances, it was decided to perform a previous analysis with different capacities of dispersion for the whole landscape, in order to verify from which distance the fragments are functionally connected. This analysis was performed in the free software Conefor Sensinode 2.6 (Saurá and Torné 2009).

Thus, the choice was to generate and analyze the Integral Index of Connectivity (IIC) for the entire landscape, selecting the option of "only overall index" in the software Conefor, considering the dispersion capacities equivalent to 50, 100, 150, 200, 250, 300, 350, 400, 450 and 500 meters.

Subsequently, the IIC results obtained were organized into an Excel spreadsheet al.ong with their respective dispersion capabilities. Next, a graphic was created to show the variation on the IIC values as a function of the analysis on different capacities of dispersion.

Processing on Conefor of Candidate Areas for Recovery

Based on the results obtained in the graphic with the different distances, the capacities of 50 and 200 meters were selected in order to identify which areas would promote a higher increase in connectivity and biological flow. For this processing, the files previously generated in the Conefor extension in ArcGIS 10.1 were inserted into the software, and the IIC was selected. Subsequently, the results were organized in an Excel worksheet and associated with the shapefile containing fragments of vegetation and the Delimited Areas for Permanent Preservation (DAPP), through the following options for analysis using the tools "Joins" and "Relates" available in the software ArcGIS 10.1.

As the purpose of the study was to prioritize the Delimited Areas for Permanent Preservation, the shapefile file with the results in two was separated, being one file with polygons of native vegetation that was represented by the number 1, and another file containing relevant polygons of the Delimited Areas for Permanent Preservation, represented by the value 0.

In this way, the values of varIIC and the fractions found were spatialized for each candidate area for restoration, the results were analyzed and the areas to be restored with a greater potential to constitute ecological corridors were prioritized (Silgueiro et al. 2017).

Then, a field was created in the table of attributes to represent the selection of approximately 17% of the total Delimited Areas for Permanent Preservation found, choosing areas with the highest values for varIICflux and varIICconnector, and also areas with the dispersion capacities of 50 and 200 meters presenting the highest values for varIICflux and varIIC connector.

RESULTS AND DISCUSSION

Candidate Areas for Recovery

The Areas for Permanent Preservation were delimited along watercourses and headwaters, according to the dimensions established by the legislation, Law 12,651 of May 25, 2012. Based on the results as shown in Table 1, the Areas for Permanent Preservation totalized 4,119.52 hectares in the study area, from which 5.74% are characterized as degraded areas and 94.26% were defined as conserved areas. When these areas are preserved, they are of great value to conserve and protect natural resources because of the roles performed by the ecosystem (Carneiro et al. 2013).

The Areas for Permanent Preservation can contain areas with a high declivity, such as in hills, mountains and water divisors, which were not delimited because of the lack of necessary data for the representation of features in the scale of 1:25.000. Artificial reservoirs were not delimited, since it was adopted the definitions based on the environmental licensing to delimit the minimum width of reservoirs in the Areas for Permanent Preservation, as approached by the legislation, Law 12 727 from October 17, 2012. The lack of access to environmental licenses for reservoirs in the Settlement Project and the absence of these licenses for the great majority made impossible the attribution of this rule in the automated model (Butturi et al. 2017).

The maintenance of Areas for Permanent Preservation is considered as one of the indispensable strategies with the purpose of maintaining the attributes of the ecosystems in proper functioning, capable of providing environmental services focused on production of water, soil conservation, carbon sequestration, biodiversity conservation and ecological balance (Jucá 2005). The Areas for Permanent Preservation are established by the legislation and are part of mitigating measures to ensure that anthropic actions does not exhaust the existing vegetation (Wammes et al. 2007).

The presence of a ciliary forest in watercourses and headwaters stands out as a key element for the conservation of these Areas for Permanent Preservation. These areas can have innumerable benefits, serving as natural corridors, assisting the flow of young seedlings, which interconnects important fragments and promotes a hydrological balance by controlling the supply of nutrients and chemicals in the watercourses (Andrade and Romero 2005). The preservation of Areas for Permanent Preservation, which contain headwaters and watercourses, benefits the physical structures of the soil, minimizing the chemical and biological contamination of water and reduces the surface runoff (Calheiros et al. 2004).

Table 1. Areas for Permanent Preservation were classified according to their conservation status as conserved and degraded areas. These areas were measured in hectares (ha) and in a relative percentage of the total area. These Areas for Permanent Preservation are localized in the Settlement Project São Pedro, in the municipality of Paranaíta, in the State of Mato Grosso, Brazil

Areas for Permanent Preservation	Area (ha)	%
Conserved	3,882.96	94.26
Degraded	236.56	5.74
Total	4,119.52	100

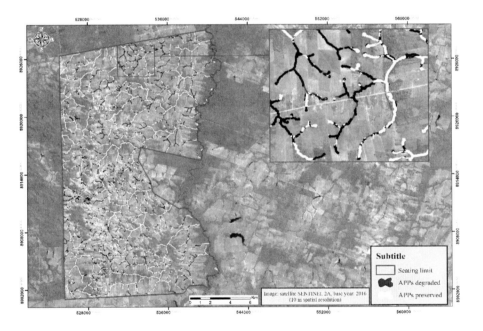

Figure 3. Areas for Permanent Preservation that were classified as conserved or degraded areas localized in the Settlement Project São Pedro.

It was noticeable that the percentage of Areas for Permanent Preservation classified as degraded areas was much lower, when compared to the conserved areas (Figure 3). This fact is due to the widths of remnants of native vegetation found in preserved Areas for Permanent Preservation, ranging from 30 to 500 meters, depending on the width of the water course. In relation to the Areas for Permanent Preservation with a degraded status, some are localized in areas of consolidated use (areas occupied before July 22, 2008), with the widths varying from 5 to 100 meters considered for restoration, depending on the size of the rural property. There is a large part of the settlement with areas of consolidated use, which may explain the considerable differences in the percentages obtained for Areas for Permanent Preservation (Butturi et al. 2017).

The analyses were performed for the municipality of Paranaíta, in the State of Mato Grosso. From Areas for Permanent Preservation, degraded areas comprised 5,580.9 hectares and preserved areas covered 35,206 hectares, considering the limits established in the Forest Code (Butturi et al. 2017). From these values, only 236.56 hectares (4.22% of degraded areas) and 3,882 hectares (11% of conserved areas) are located in the Settlement Project São Pedro.

The use of the automated model presented satisfactory results, facilitating the acquisition of the Areas for Permanent Preservation in a quick and reliable way. This fact is related to the quality of the databases used, as well as the level of details. The bases were elaborated in a standardized way by the Instituto Centro de Vida, with the absence of overlaps and verification of topological errors. With the automatic delimitation of the Areas for Permanent Preservation, the subjectivity of the process was eliminated.

Through the registration in the Environmental Rural Register, which includes the survey of documentary and georeferenced information, the quantity and distribution of the environmental liabilities are known to be associated with the Areas for Permanent Preservation from rural properties. From the results, the owner of the property is required to restore the native vegetation in the Areas for Permanent Preservation that present the original vegetation suppressed. However, the stage of the environmental regularization in Settlement Project has slowly evolved, despite six years after the implementation of the New Forest Code.

In order to perform the environmental regularization in an official manner, it is necessary to validate the Environmental Rural Register of the property by the competent environmental agency and right after the owner of the rural property needs to prepare and execute the Recovery Plan for areas in a degraded and/or in an altered status (PRADA).

The great difficulty to perform the restoration of the Areas for Permanent Preservation is related to the mode of rural production and the environmental conservation, since the rural owners perceive them as an economic and cultural barrier to the conservation and recovery of these areas, considering as a legal provision for punishment and not of land management (Kageyama et al. 2007).

The main strategies with the purpose of recovering Areas for Permanent Preservation should be adopted in partnerships, which in other words mean that instruments of command and control of the legislation should work together with economic instruments that have resulted in the conservation and preservation of these areas (Ranieri 2014). Before implementing any recovery strategy, it must consider the extent of these legal devices in small rural properties (Kageyama et al. 2007).

The application of geotechnology has proved to be a viable alternative to reduce the time spent to map the Areas for Permanent Preservation, and can assist by the anticipation of the environmental regularization process.

Analysis of Fragments and the Euclidian Distance

Based on the results obtained through the analysis using the tool for the calculus of the Euclidean Distance, it was possible to perform the preliminary analysis of the landscape, allowing to limit the number of polygons to avoid problems during data processing. The results indicate that the landscape is functionally connected from 500 meters, discarding the analyses with dispersion capacities greater than 500 meters.

Through the interpretation of the map, it was found that the largest distance between fragments is close to 500 meters. This distance can be quite variable and is dependent on the size of the area under analysis. In the studies performed in the municipality of Alta Floresta, in the State of Mato Grosso, with a limit distance of 1,000 meters due to the land extension, it was necessary a bigger limit distance in comparison to the limit distance considered in the Settlement Project (Tambosi et al. 2016). The fragments of native vegetation and the values of distances in meters are showed on the map represented in Figure 4.

It was verified on the map that a great quantity of the fragments of native vegetation corresponds to small classes of size, which may result in aspects directed to degradation and a poor state of conservation.

Regardless of the size of fragments, both show a great relevance in the landscape, which means that larger fragments act in the maintenance of biodiversity and large-scale ecological processes, and smaller fragments are connection elements between large areas (Forman and Godron 1986).

Figure 4. Distance maps obtained by using fragments of native vegetation.

Prioritization of Areas for Permanent Preservation

The maintenance of populations in small forest fragments is potentialized through the existence of Areas for Permanent Preservation, due to the fact that they act in the connection of fragments and in the expansion of the forest cover, which benefits life cycle in these habitats and favoring the creation of new environments due to the fact that each fragment presents exclusive characteristics (Carneiro et al. 2013).

Analysis of Landscape for Different Dispersion Abilities

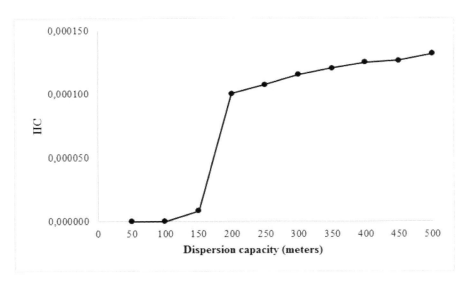

Figure 5. Values for Integral Index of Connectivity (IIC) that were obtained based on different dispersion abilities.

Due to the large number and size of the landscape fragments, high IIC values were obtained because their calculus multiplies the values for connected areas from fragment pairs. In order to facilitate data interpretation and reduce the values, it was opted to divide each result by the settlement area squared (348,948,499.11 meters2). Thus, the values ranged from 0 to 1 for all dispersion capacities. Thus, the IIC value increases as the landscape becomes more connected.

Based on the results (Figure 5), it is possible to observe that the greatest variations in relation to the dispersion capacity are between distances of 150 and 200 meters. In this context, the isolation between the fragments can reflect in relevant differences for species that move in smaller capacities than 200 meters. From 350 meters there are no large variations, which suggests that organisms are already functionally connected.

The definition of the dispersion capability can be performed based on the information about the organisms present in the study area, on a species of interest and on the situation of landscape fragmentation. Thus, this avoids the application of low capacities of movement in very fragmented landscapes, as well as the use of superior capacities to the average isolation of the fragments before the restoration.

Based on the graph from Figure 5, the landscape analysis were performed with the purpose to identify the priority areas for restoration, with the choice to consider the dispersion capacities of 50 and 200 meters and verify the importance of the Delimited Areas for Permanent Preservation in relation to the biological flow and landscape connectivity.

Analysis of Landscape and Prioritization of Delimited Areas for Permanent Preservation for Recovery

The results found for the prioritization of the Delimited Areas for Permanent Preservation were performed based on the landscape connectivity, biological flow and when considering both traits for the delimited areas. Initially, it was decided to prioritize around 20 hectares for each of the observed variables. Areas with high values for the flow as well as connectivity were outlined and recognized separately. The analyzes were performed for the dispersion capacities of 50 and 200 meters, based on analysis of the graph containing the dispersion capacity (Figure 5).

From the Areas for Permanent Preservation, a total of 236.36 hectares are degraded areas with both dispersion abilities. Around 40 hectares were selected, which represents 17.1% of the Delimited Areas for Permanent

Prioritization of Areas for Permanent Preservation 135

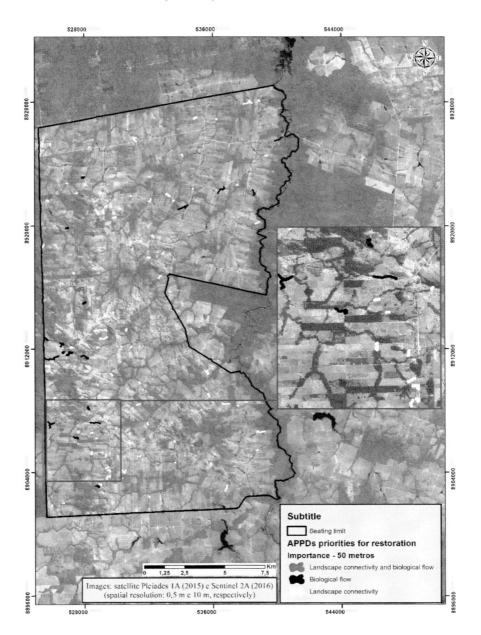

Figure 6. Prioritized Delimited Areas for Permanent Preservation with the purpose of forest recovery in the Settlement Project São Pedro, in the municipality of Paranaíta, in the State of Mato Grosso, with the intention of building ecological corridors and the recovery success on a movement capacity of 50 meters.

Preservation to be recovered, from which around 6 hectares with the dispersion capacity of 50 meters were outlined as priority areas for both connectivity and the flow (Table 2).

For the analyzes performed considering the dispersion capacity of 50 meters, the areas outlined are relevant for both the landscape connectivity as well as the biological flow, and they are considered as priorities areas for restoration as shown in grey color in Figure 6. These areas have the potential to create functional connections between two isolated fragments, as well as to maximize the biological flow, characterizing themselves as possible ecological corridors.

Table 2. Distribution of the prioritized areas from the Delimited Areas for Permanent Preservation with the purpose of forest recovery, according to the relevant requirements for 50 meters of movement

Relevance of the Prioritized Areas from the Delimited Areas for Permanent Preservation	Area (ha)	%
Landscape Connectivity	17	7.3
Biological Flow	17	7.3
Landscape Connectivity and biological flow	6	2.5
Total	40	17.1

The most important areas considered only for landscape connectivity totalized 17 hectares, which represents approximately 7.3% of the total area to be restored, as shown in white color in Figure 6.

Connectivity is related to the landscape capacity to facilitate biological flow between landscape fragments (Taylor et al. 2003). Landscapes with a high degree of connectivity allow organisms to be dispersed between the elements presented in the landscape, resulting in a great biological flow in the maintenance of the high diversity and reduction of local extinction risks (Pardini et al. 2010).

Most of the rural properties are in areas with a long history of anthropic occupation, which results in remnants of isolated vegetation presenting degradation in a great part of these rural properties. In this context, the

Prioritization of Areas for Permanent Preservation 137

configuration of the landscape is of great value to promote the maintenance of biodiversity (Andrén 1994) and, therefore, landscapes with closer and connected fragments result in a greater potential to shelter biodiversity (Pardini et al. 2010). Thus, forest areas that are isolated in smaller fragments may not present original species due to the natural distribution of spots for species in response to the environmental heterogeneity (Scariot et al. 2005).

Among innumerous anthropic disturbances, habitat loss and fragmentation are known as the main causes for the decline of species and populations (Fahrig 2003). The reduction of the available habitat is considered the major current cause for biodiversity decline, resulting in a decreased connectivity, increased edge effect and, consequently, in environmental degradation (Ewers and Didham 2006).

Besides the loss of species as a consequence of fragmentation processes, it can occur a species migration to vegetation fragments, serving as a refuge of species (Lovejoy 1980). Species that require a wide or specialized habitat are more susceptible to the effects of fragmentation (Turner 1996).

The Delimited Areas for Permanent Preservation with the highest potential of biological flow are shown in black color in Figure 6 and totalized 17 hectares. These areas are determinant for the success of forest restoration, since they present a potential for the recolonization of fauna and flora species.

The presence of a flow of organisms among the vegetation fragments allows the restored areas to be colonized by flora species brought by animal seed dispersal. The colonization of areas under restoration by fauna species assists pollination and seed dispersion among landscape fragments (Ghazoul 2005).

For the analyzes performed for the dispersion capacity of 200 meters (Table 3), 10.92 hectares of priority areas were obtained for the landscape connectivity as well as for the biological flow, which is approximately the double in comparison to the results obtained for a movement of 50 meters.

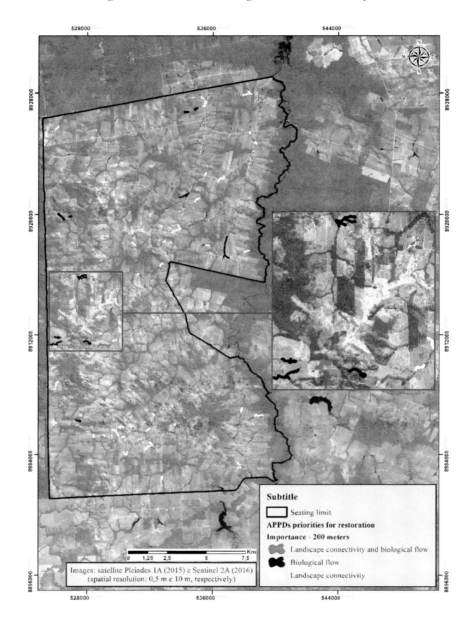

Figure 7. Prioritized Delimited Areas for Permanent Preservation for forest recovery in the Settlement Project São Pedro, in the municipality of Paranaíta, in the State of Mato Grosso, with the purpose to build ecological corridors and the success of recovery actions for the movement of 200 meters.

Thus, it is possible to obtain more relevant areas for both variables when there is a greater capacity of movement, which facilitate the connection between the fragments and biological flow.

There were around 6.29% of relevant areas for landscape connectivity and 6.37% of priority areas for the biological flow, lower values than the capacity of movement of 50 meters, which accounts for more important areas in relation to the observed variables as it is showed in Figure 7. The percentage chosen to prioritize the areas corresponds to only 17.36% of the total areas to be restored in the Settlement Project São Pedro.

In this study, it is verified that initiatives and projects prioritize important areas for connectivity and biological flow, being more viable to use the distance of 200 meters as the dispersion capacity, due to the fact of obtaining more areas that allow the establishment of ecological corridors.

Ecological corridors are characterized as one of the most promising strategies for effective regional plans for the conservation and preservation of flora and fauna species (Valeri and Senô 2003), which can benefit innumerous factors, as well as presenting appropriate conditions for species survival and reproduction, acting as a disperser to facilitate the recolonization of species by reduced populations and as a water filter that improves water quality. The corridors aim to increase the survival probability of individuals and genes between populations in a long term, which results in the maintenance of ecological processes, since there is a connection of small fragments to large protected areas (Ayres et al. 2005).

The application of landscape metrics allowed to identify the areas with the greatest connectivity between forest fragments and the re-establishment of fauna and flora species, as well as the definition of priority areas with the purpose to construct ecological corridors. Therefore, after analyzing the IIC values that were calculated, some requirements were defined to select priority areas for restoration with the purpose to optimize future efforts for the assertive implementation and cost reduction in restorations performed in these areas.

Among the traits considered of great importance for the restoration success, it is noted the proximity of forest fragments of the same phytogeographic unit, the permeability of the matrix, which allows the arrival of natural propagules and dispersal of organisms with the presence of a bank containing viable seeds from native species and a degree of soil compaction (Holl 1999).

Table 3. Distribution of Delimited Areas for Permanent Preservation for forest recovery, according to the relevance of requirements for a movement of 200 meters

Relevance of Priority Areas from the Delimited Areas for Permanent Preservation	Area (ha)	%
Landscape of connectivity	14.60	6.29
Biological Flow	14.78	6.37
Landscape Connectivity and Biological Flow	10.92	4.70
Total	40.2	17.36

The definition of the method to be applied as the most adequate restoration action to each situation is considered as the key to success for restoration initiatives, since it results in the reduction of costs and time (Nave et al. 2015). It is necessary to consider the history of the areas to be restored, which is obtained through a time series analysis, as well as the current situation and the information provided by the owners. In a same region or property, different restoration methodologies should be used, according to the specific characteristics identified in the zone area (Rodrigues et al. 2009).

In this context, field visits are of great value after obtaining the first diagnoses of the place, which allows a more accurate survey and verify cost implications, area size and the potential benefits for connectivity. The criteria related to the quantity of selected areas are linked to the purposes of the selection process and resources available for restoration (Silgueiro et al. 2017).

An important criterion used in analyzes is the crossing of rural properties with the priority Delimited Areas for Permanent Preservation found in the

Settlement Project, which allows the verification of properties that can be involved in restoration projects (Silgueiro et al. 2017). It should be noted that there is no unique rule to be followed, and the analysis of the results must be performed in a specific way for each case, with the purpose to identify which one is the most suitable for the purpose of the analysis.

Restoration actions with the aim to adequate rural properties are constantly distributed over years and decades, due to the great quantity of rural properties that have significant environmental liabilities, according to the requirements of the New Forest Code (Igari et al. 2009; Rodrigues et al. 2009). Therefore, it is necessary to be cautious when establishing restoration goals for a short term, because the results obtained after analysis are dependent on the area conditions and the climate characteristics of the site.

In Settlements Projects there is a great difficulty to work with the conservation of natural resources, since the occupation dynamics and land use, the socioeconomic aspects have always been prioritized and the environmental aspect have been seen as an obstacle.

A process of agricultural land reform should open perspectives with the purpose of recovering degraded areas and the environmental care by the settlers, since this will make the sustainability of their production system. The action strategies for the construction of an environmental program mainly involve plans and land organization of the Settlement Project, as well as the adoption of differentiated agricultural practices, which involves the use of agroecology (Bertolini and Carneiro 2007).

There are great challenges for the process of agricultural land reform, since the form of occupation in the Settlements has been marked by land reconcentration. The role of the State on the viability of the Settlements Projects is another challenging fact that needs to be reconstructed, not only as a social policy, but also as a policy for socioeconomic development, with perspectives for the recovery of degraded areas (Bertolini and Carneiro 2007).

CONCLUSION

The use of geotechnologies has proved to be such as a viable alternative for obtaining Areas for Permanent Preservation and for making decisions about recovery actions, helping in the process of environmental regularization.

The automated model for the diagnosis of Delimited Areas for Permanent Preservation was satisfactory, making possible to obtain reliable data in a short-term. This fact was due to the quality of the databases used, which directly influenced the acquisition of results.

Landscape analysis demonstrates the relevance of plans aiming recovery actions of forest. The prioritization of the Delimited Areas for Permanent Preservation to be restored is based on values of importance that were attributed for landscape connectivity and biological flow, which can make viable the constitution of ecological corridors, promoting the maintenance of ecological processes and the preservation of fauna and flora species, as well as the reduction of costs associated with recovery actions.

REFERENCES

Andrade LMS de, Romero MAB (2005) A importância das áreas ambientalmente protegidas nas cidades. XI Encontro Nacional de Pós-Graduação e Pesquisa em Planejamento Urbano e Regional – ANPUR. Salvador [The importance of environmentally protected areas in cities. XI National Meeting of Postgraduate and Research in Urban and Regional Planning - ANPUR. Salvador].

Andrén H (1994) Effects of habitat fragmentation on birds and mammals in landscapes with different proportions of suitable habitat: a review. *Oikos,* pp 355-366.

Ayres JM et al. (2005) Os corredores ecológicos das florestas tropicais do Brasil. Belém, PA: Sociedade Civil Mamirauá. BRASIL. Decreto Federal Nº 4340, de 22 de agosto de 2002 [The ecological corridors of

Prioritization of Areas for Permanent Preservation 143

the tropical forests of Brazil. Belém, PA: Mamirauá Civil Society. BRAZIL. Federal Decree No. 4340 of August 22, 2002].

Bennett AF (2003) Linkages in the landscape: the role of corridors and connectivity in wildlife conservation. Gland, Switzerland and Cambridge, United Kingdom: The World Conservation Union (IUCN) Forest Conservation Programme, 2 ed.

Bernasconi P et al. (2009) Avaliação Ambiental Integrada: Território Portal da Amazônia. Instituto Centro de Vida (ICV), Alta Floresta [Integrated Environmental Assessment: Territory Portal da Amazônia. Instituto Centro de Vida (ICV), Alta Floresta].

Bertolini VA, Carneiro FF (2007) Considerações sobre o planejamento espacial e a organização da moradia dos assentamentos de reforma agrária no DF e entorno. *Libertas,* edição especial, pp 202-226 [Considerations on the spatial planning and the organization of housing for agrarian reform settlements in DF and surroundings. *Libertas,* special edition, pp. 202-226].

Brasil. Lei N° 12.651 de 25 de maio de 2012. Dispõe sobre a proteção da vegetação nativa; altera as Leis nos 6.938, de 31 de agosto de 1981, 9.393, de 19 de dezembro de 1996, e 11.428, de 22 de dezembro de 2006; revoga as Leis nos 4.771, de 15 de setembro de 1965, e 7.754, de 14 de abril de 1989, e a Medida Provisória no 2.166-67, de 24 de agosto de 2001; e dá outras providências. Diário Oficial da União, Brasília-DF, 28 mai. 2012. http://www.planalto.gov.br/ccivil_03/_ato2011-2014/2012/lei/ 112651.htm [Provides for the protection of native vegetation; amends Laws 6,938 of August 31, 1981, 9,393 of December 19, 1996, and 11,428 of December 22, 2006; revokes Laws Nos. 4,771, September 15, 1965, and 7,754, April 14, 1989, and Provisional Measure No. 2,166-67 of August 24, 2001; and makes other arrangements. Official Journal of the Union, Brasília-DF, May 28. 2012].

Brasil. Lei N° 4.504 de 30 de novembro de 1964. Dispõe sobre o Estatuto da Terra, e dá outras providências. Diário Oficial da União, Brasília-DF, 30 nov. 1964. http://www.planalto.gov.br/ccivil_03/leis/ L4504.htm [It deals with the Statute of the Earth, and gives other measures. Official

Journal of the Union, Brasília-DF, Nov. 30 1964 http://www.planalto.gov.br/ccivil_03/leis/L4504.htm].

Buchanan GM et al. (2009) Delivering a global, terrestrial biodiversity observation system through remote sensing. *Conservation Biology*, pp 499-502.

Bunn AG et al. (2000) Landscape connectivity: a conservation application of graph theory. *Journal of Environmental Management*, pp 266-278.

Butturi W et al. (2017) Modelo para delimitação automática de áreas de preservação permanente conforme o Novo Código Florestal: aplicação em três municípios no Bioma Amazônia em Mato Grosso. In: *Simpósio Brasileiro de Sensoriamento Remoto*, 18. (SBSR), Santos. Anais... São José dos Campos: INPE, pp 910-917. Internet. ISBN 978-85-17-00088-1. IBI: 8JMKD3MGP6W34M/3PS4FQP. http://urlib.net/ 8JMKD3MGP6W34M/3PS4FQP [Model for automatic delimitation of permanent preservation areas according to the New Forest Code: application in three municipalities in the Amazon Biome in Mato Grosso. In: *Brazilian Symposium on Remote Sensing*, 18. (SBSR), Santos. Anais ... São José dos Campos: INPE, pp 910-917. Internet. ISBN 978-85-17-00088-1. IBI: 8JMKD3MGP6W34M / 3PS4FQP. http://urlib.net/8JMKD3MGP6W34M/3PS4FQP].

Calheiros RO et al. (2004) Preservação e recuperação das nascentes (de água e de vida). Comitê das Bacias Hidrográficas dos Rios Piracicaba, Capivarí e Jundiaí – CTRN. Piracicaba [Preservation and recovery of springs (water and life). Committee on the Hydrographic Basins of Piracicaba, Capivarí and Jundiaí Rivers - CTRN. Piracicaba].

Carneiro BM et al. (2013) Perspectivas de conexão entre fragmentos florestais do corredor ecológico Burarama-Pacotuba-Cafundó, na mata atlântica do Espírito Santo, através de recomposição de áreas de proteção permanente de cursos d'água. *Natureza Online*, pp 20-28 [Perspectives of connection between forest fragments of the ecological corridor of Burarama-Pacotuba-Cafundó, in the Atlantic Forest of Espírito Santo, through the recomposition of areas of permanent protection of watercourses. *Nature Online*, pp. 20-28].

Prioritization of Areas for Permanent Preservation 145

Crouzeilles R, Curran M (2016) Which landscape size best predicts the influence of forest cover on restoration success? A global meta-analysis on the scale of effect. *Journal of Applied Ecology*, pp 440-448.

Cruz ZQ da, Ribeiro GP (2008) Ensaios de segmentação e classificação digital de imagens cbers utilizando o sistema spring em uma unidade de conservação ambiental estudo de caso: parque nacional da serra dos órgãos (parnaso). In: *II Simpósio Brasileiro de Ciências Geodésicas e Tecnologias da Geoinformação*, 9. Recife. Anais... Recife: UFPe. https://www.ufpe.br/cgtg/SIMGEOII_CD/Organizado/sens_foto/023. pdf [Segmentation and digital classification of cbers images using the spring system in an environmental conservation unit case study: national park of the Serra dos Organias (Parnassus). In: 2nd *Brazilian Symposium on Geodetic Sciences and Geoinformation Technologies*, 9. Recife. Anais ... Recife: UFPe. https://www.ufpe.br/cgtg/ SIMGEOII_CD/Organizado/sens_foto/023.Pdf].

Estevam LS, Pereira AS (2015) As áreas de preservação permanente a luz do novo código florestal. In: *Simpósio Brasileiro de Sensoriamento Remoto,* 17. (SBSR), 2015, João Pessoa, PB, Brasil. Anais... São José dos Campos: INPE, pp 2301-2308. Online. ISBN: 978-85-17-0076-8. http://www.dsr.inpe.br/sbsr2015/files/p0470.pdf [The areas of permanent preservation light the new forest code. In: *Brazilian Symposium on Remote Sensing*, 17. (SBSR), 2015, João Pessoa, PB, Brazil. Anais ... São José dos Campos: INPE, pp 2301-2308. Online. ISBN: 978-85-17-0076-8. http://www.dsr.inpe.br/sbsr2015/files/ p0470.pdf].

Ewers RM, Didham RK (2006) Confounding factors in the detection of species responses to habitat fragmentation. *Biological Review*, pp 117-142.

Fahrig L (2003) Effects of habitat fragmentation on biodiversity. *Annual Review of Ecology, Evolution and Systematics*, pp 487-515.

Forman RTT, Grodon R (1986) *Landscape Ecology*. John Wiley e Sons, Inc. New York.

Ghazoul J (2005) Pollen and seed dispersal among dispersed plants. *Biological Reviews*, pp 413-443.

146   *E. Brugnara, V. de Freitas Silgueiro and J. C. Wojciechowski*

Hoffmann MRM (2015) Restauração florestal mecanizada: semeadura direta sobre palhada. Instituto Centro de Vida (ICV), Alta Floresta. http://www.icv.org.br/wpcontent/uploads/2015/11/Restauracao_florest al_novembro2015.pdf [Mechanized forest restoration: direct sowing on straw. Instituto Centro de Vida (ICV), Alta Floresta. http://www.icv.org.br/wpcontent/uploads/2015/11/Restauracao_florest al_novembro2015.pdf].

Holl KD (1999) Tropical forest recovery and restoration. *TRENDS in Ecology and Evolution,* pp 378-379.

Horning N et al. (2010) *Remote Sensing for Ecology and Conservation: A Handbook of Techniques.* Oxford University Press, New York.

Instituto Brasileiro de Geografia e Estatística - IBGE (2012) Manual Técnico da Vegetação Brasileira Sistema Fitogeográfico Inventário das formações florestais e campestres Técnicas e manejo de coleções botânicas Procedimentos para mapeamentos, Rio de Janeiro [Technical Manual of Brazilian Vegetation Phytogeographic System Inventory of forest and field formations Techniques and management of botanical collections Procedures for mapping, Rio de Janeiro].

Igari AT, Tambosi LR, Pivello VR (2009) Agribusiness opportunity costs and environmental legal protection: investigating trade-off on hotspot preservation in the state of São Paulo, Brazil. *Environmental Management,* pp 346-355.

Instituto Centro de Vida - ICV (2010) Calendário da Bacia do Alto Paraguai. http://www.icv.org.br/wpcontent/uploads/2013/08/67736 calendario_da_bacia_do_alto_paraguai_pdf. [Calendar of the Upper Paraguay Basin. http://www.icv.org.br/wpcontent/uploads/2013/ 08/67736calendar_of_the_bath_of_the_paraguai_pdf.].

Instituto Centro de Vida - ICV (2016) Análise sobre regularização ambiental da agricultura familiar em Mato Grosso. http://www.icv.org.br/2016/11/28/icv-divulga-analise-sobre-a-regularizacao-ambiental-da-agricultura-familiar-em-mato-grosso/ [Analysis of environmental regularization of family agriculture in Mato Grosso. http://www.icv.org.br/2016/11/28/icv-divulga-analysis-sobre-a-regularizaca-ambiental-of-agricultura-familiar-in-ma-grosso/].

Instituto Nacional de Colonização e Reforma Agraria - INCRA (2016) http://www.incra.gov.br/search/node/Modulo%20fiscal.

Jesus EN (2013) Avaliação dos fragmentos florestais da bacia hidrográfica do rio poxim (Sergipe-Brasil) para fins de restauração ecológica. Dissertação, Universidade Federal de Sergipe [Evaluation of the forest fragments of the poxim river basin (Sergipe-Brazil) for the purpose of ecological restoration. Dissertation, Federal University of Sergipe].

Jucá FT (2007) Marcos legais sobre reserva legal e áreas de preservação permanente: uma estratégia para conservação dos recursos naturais. Monografia, Universidade Rural do Rio de Janeiro [Legal frameworks on legal reserve and permanent preservation areas: a strategy for the conservation of natural resources. Monograph, Rural University of Rio de Janeiro].

Kageyama PY et al. (2007) Carta de Piracicaba: I fórum sobre APP e RL na paisagem e propriedade rural. Piracicaba: ESALQ/USP [Carta de Piracicaba: I forum on APP and RL in landscape and rural property. Piracicaba: ESALQ / USP].

Lovejoy TE Foreword. In: Soulé ME, Wilcox BA (1980) *Conservation biology: an evolutionary-ecological perspective*. Sunderland: Sinauer Associates, pp 5-9.

Metzger JP (1999) Estrutura da paisagem e fragmentação: análise bibliográfica. *Anais da Academia Brasileira de Ciências*, v. 71, pp 445-462 [Landscape structure and fragmentation: bibliographic analysis. *Annals of the Brazilian Academy of Sciences*, v. 71, pp 445-462].

Muchailh MC et al. (2010) Metodologia de planejamento de paisagens fragmentadas visando a formação de corredores ecológicos. *Floresta,* pp 147-162 [Methodology of planning fragmented landscapes aiming at the formation of ecological corridors. *Forest*, pp. 147-162].

Nascimento Soares JL (2008) A organização territorial de assentamentos rurais para atender a legislação ambiental na Amazônia. *Revista de Geografia Agrária*, pp 143-155 [The territorial organization of rural settlements to meet environmental legislation in the Amazon. *Revista de Geografia Agrária,* pp. 143-155].

Nave A et al. (2015) Manual de restauração ecológica: técnicos e produtores rurais no extremo sul da Bahia. *Bioflora Tecnologia de Restauração.* Piracicaba [Manual of ecological restoration: technicians and rural producers in the extreme south of Bahia. *Bioflora Restoration Technology.* Piracicaba].

Oliveira ALA, Bergamasco SMPP (2014) Fortalecimento da agricultura familiar: uma análise do Pronaf no Projeto de Assentamento São Pedro, Paranaíta, MT. *Retratos de Assentamentos*, n. 1 [Strengthening family agriculture: an analysis of Pronaf in the Settlement Project São Pedro, Paranaíta, MT. *Portraits of Settlements*, n. 1].

Paese A (2002) A utilização de modelos para a análise da paisagem na região nordeste do estado de São Paulo. Tese, Universidade Federal de São Carlos [The use of models for landscape analysis in the northeastern region of the state of. Thesis, Federal University of São Carlos].

Paese A et al. (2012) Conservação da biodiversidade com SIG. 1. ed. São Paulo: Oficina de Textos, v.1, pp 240 [Biodiversity conservation with GIS. 1. ed. São Paulo: Oficina de Textos, v.1, pp 240].

Pardini R et al. (2010) Beyond the fragmentation threshold hypothesis: regime shifts in biodiversity across fragmented landscapes. *Plos One,* pp 1366.

Pettorelli N et al. (2014) Satellite remote sensing, biodiversity research and conservation of the future. *Philos. Trans. R. Soc. B* 369, 20130190.

Proulx SR et al. (2005) Network thinking in ecology and evolution. *TRENDS in Ecology and Evolution*, pp 345-356.

Ranieri VEL (2004) Reservas Legais: critérios para localização e aspectos de gestão. Tese, Universidade de São Paulo [Legal Reserves: criteria for location and management aspects. Thesis, University of São Paulo].

Rezende RA (2011) Fragmentação da Flora Nativa como Instrumento de Análise da Sustentabilidade Ecológica de Áreas Protegidas – Espinhaço Sul (MG). Tese, Universidade Federal de Ouro Preto [Fragmentation of the Native Flora as an Instrument for Analysis of the Ecological Sustainability of Protected Areas - Espinhaço Sul (MG). Thesis, Federal University of Ouro Preto].

Rodrigues RR et al. (2009) PACTO pela restauração da Mata Atlântica: referencial dos conceitos e ações de restauração florestal. São Paulo:

LERF/ESALQ, Instituto Bioatlântica. http://www.pactomataatlantica. org.br/pdf/referencial-teorico.pdf [PACT for the restoration of the Atlantic Forest: reference of the concepts and actions of forest restoration. São Paulo: LERF / ESALQ, Bioatlantic Institute. http:// www.pactomataatlantica.org.br/pdf/referencial-teorico.pdf].

Saura S, Rubio L (2010) A common currency for the different ways in which patches and links can contribute to habitat availability and connectivity in the landscape. *Ecography,* pp 523-537.

Saura S, Torné J (2009) Conefor Sensinode 2.2: a software package for quantifying the importance of habitat patches for landscape connectivity. *Environmental Modelling & Software*, pp 135-139.

Scariot A et al. (2005) Cerrado: ecologia biodiversidade e conservação. In: Felfili JM, Silva JCS, Scariot A. (ed.). Biodiversidade, ecologia e conservação do Cerrado: avanços no conhecimento. Brasília: Ministério do Meio Ambiente, cap. 1, pp 27-44 [Closed: ecology biodiversity and conservation. In: Felfili JM, Silva JCS, Scariot A. (ed.). Biodiversity, ecology and conservation of the Cerrado: advances in knowledge. Brasília: Ministry of the Environment, chap. 1, pp. 27-44].

Silgueiro VF et al. (2017) Identificação de áreas de preservação permanente prioritárias para restauração florestal visando a constituição de corredores ecológicos nos municípios de Alta Floresta, Carlinda e Paranaíta em Mato Grosso. In: *Simpósio Brasileiro de Sensoriamento Remoto,* 18. (SBSR), Santos. Anais... São José dos Campos: INPE, pp 1329-1336. Internet. ISBN 978-85-17-00088-1. IBI:8JMKD-3MGP6W34M/3PS4GGQ. http://urlib.net/8JMKD3MGP6W34M-/3PS4GGQ [Identification of priority permanent preservation areas for forest restoration aiming at the constitution of ecological corridors in the municipalities of Alta Floresta, Carlinda and Paranaíta in Mato Grosso. In: *Brazilian Symposium on Remote Sensing,* 18. (SBSR), Santos. Anais ... São José dos Campos: INPE, pp 1329-1336. Internet. ISBN 978-85-17-00088-1. IBI: 8JMKD3MGP6W34M / 3PS4GGQ. http://urlib.net/ 8JMKD3MGP6W34M/3PS4GGQ].

Silgueiro VF et al. (2015) Uso do balão para imageamento de alta resolução a baixo custo: aplicação para o monitoramento de áreas de restauração florestal. In: *Simpósio Brasileiro de Sensoriamento Remoto*, 17. (SBSR), João Pessoa, PB, Brasil. Anais… São José dos Campos: INPE, pp 4017-4024. Online. ISBN: 978-85-17-0076-8. http://www.dsr. inpe.br/sbsr2015/files/p0796.pdf [Use of the balloon for high resolution imaging at low cost: application for the monitoring of forest restoration areas. In: *Brazilian Symposium on Remote Sensing*, 17. (SBSR), João Pessoa, PB, Brazil. Anais ... São José dos Campos: INPE, pp 4017-4024. Online. ISBN: 978-85-17-0076-8. http://www.dsr.inpe. br/sbsr2015/files/p0796.pdf].

Sistema Brasileiro de Classificação de Solos - SIBCS (2013) Empresa Brasileira de Pesquisa Agropecuária - EMBRAPA. 3.ed. Brasília [Brazilian Agricultural Research Corporation - EMBRAPA. 3.ed. Brasília].

Tambosi LR (2014) Estratégias espaciais baseadas em ecologia de paisagens para a otimização dos esforços de restauração. Tese, Universidade de São Paulo [Spatial strategies based on landscape ecology for the optimization of restoration efforts. Thesis, University of São Paulo].

Tambosi LR et al. (2016) Uso das geotecnologias para o planejamento espacial e monitoramento da restauração florestal em áreas de preservação permanente degradas (APPDs): Experiências nos municípios de Alta Floresta, Carlinda e Paranaíta em Mato Grosso. https://www.icv.org.br/wp-content/uploads/2016/12/Monitoramento _APPDs_web.pdf [Use of geotechnologies for spatial planning and monitoring of forest restoration in degraded permanent preservation areas (APPDs): Experiments in the municipalities of Alta Floresta, Carlinda and Paranaíta in Mato Grosso. http://www.icv.org.br/wp-content/uploads/2016/12/Monitoramento_ APPDs_web.pdf].

Taylor PD et al. (1993) Connectivity is a vital element os landscape structure. *Oikos*, pp 571-573.

Turner IM (1996) Species loss in fragments of tropical rain forest: a review of the evidence. *Journal of Applied Ecology,* pp 200-209.

Urban D, Keitt T (2001) Landscape connectivity: a graph-theoretic perspective. *Ecology,* pp 1205-1218.

Valeri SV, Senô MAAF (2003) A importância dos corredores ecológicos para a fauna e a sustentabilidade de remanescentes florestais. http://www.saoluis.br/revistajuridica/arquivos/005.pdf [The importance of ecological corridors for fauna and the sustainability of forest remnants. http://www.saoluis.br/revistajuridica/arquivos/ 005.pdf].

Wammes EVS et al. (2007) Importância ambiental das áreas de preservação permanente e sua quantificação na microbacia hidrográfica da Sanga Mineira do município de Mercedes - PR. *Revista Brasileira de Agroecologia,* pp 1408-1411 [Environmental importance of permanent preservation areas and their quantification in the Sanga Mineira watershed of the municipality of Mercedes - PR. *Brazilian Journal of Agroecology*, pp. 1408-1411].

Wilcove DS et al. (1986) Habitat fragmentation in the temperate zone. In: *Conservation Biology*, ed. ME Soule, Sunderland, MA: Sinauer, pp 237–256.

Wimberly MC (2006) Species Dynamics in Disturbed Landscapes: When does a Shifting Habitat Mosaic Enhance Connectivity? *Landscape Ecology,* pp 35-46.

In: Forest Conservation
Editor: Pedro V. Eisenlohr

ISBN: 978-1-53614-559-5
© 2019 Nova Science Publishers, Inc.

Chapter 5

CONSERVATION OF ALEPPO PINE FOREST FOR POST FLOOD AND FIRE PLANTINGS

Abdelaziz Ayari[*]
Institut National de Recherches en Génie Rural, Eaux et Forêts
(INRGREF), Département d'Ecologie Forestière, Tunis, Tunisia

INTRODUCTION

Worldwide, Aleppo pine forests cover about 3.5 million hectares, where 2.5 million hectares occupied the Mediterranean region (Rigolot et al. 2012). Its geographical area ranges from Portugal to minor countries of Asia from west-east, and from southern France to Maghreb Arabic countries for the displacement of the north to south coast of Mediterranean Sea (Krugman and Jenkinson 1974; Fady et al. 2003). The species forests area within the Maghreb countries is about 1.2 million hectares (Mezali, 2003). Fady et al. (2003) and Antipolis (2001) restricted the species with 1,586,000 hectares in Europe and the northern coast of the Mediterranean Sea. Recent published reports (Hall et al. 1997, Gibbs et al. 1999, Spencer 2001 and Hokche et al.

---

[*] Corresponding Author Email: abdelazizayari@yahoo.fr.

2008) found small and reduced species forest plantations in other non-Mediterranean zones (USA, Venezuela, Argentina, Australia and South Africa).

The Aleppo pine is the main forest tree species within the Maghreb countries, where its area ranges from the wet bioclimatic zone in the north (sub-humid) to the dry bioclimatic zone in the south (semi-arid). Ayari and Khouja (2014) reported that the semi-arid zone is conforming to the species bioclimatic region. More than 50% of the species forest cover within Tunisia, where its main bioclimatic zone is the upper semiarid (DGF 1995 and 2010). Ayari et al. (2011c) published that the stand species density is ranging between 250 trees/ha to about 2200 trees/ha. Moreover, Ayari et al. (2011a and 2011b) showed that low Aleppo pine densities are located within open and degraded stand mixed regularly by varies species. Specialists' efforts have also published that its forest growing area show an approximate of thirteen significant accompanying species groups within the whole forest in Tunisia.

Likewise, the Aleppo pine trees are characterized by massive yearly seed production (Ayari et al. 2012c). The species utilization is for several reasons, such as reforestation, anti-erosion and edible seeds, in addition to wood production (Ayari et al. 2012b; Ayari and Khouja 2014). The species wood quality is poor but significant in uses, which are motivated to decoration, papermaking and resin production (Ayari 2012). Equally, its bark is so rich in tannin (preparation of skins) largely observed for older Aleppo pine trees (Naveh 1974; Chakroun 1986). The first aim of this chapter is to assess species fruitation for a set of Tunisian provenances planted in a coastal Aleppo pine stand, which managed for seed and cone production under similar ecological conditions. The second one is to help foresters and stakeholders to define forest zone within their countries for in-situ gene conservation of forest species having better rates as native plant materials for post-flood or post-fire plantings.

MATERIAL AND METHODS

Study Site and Natural Vegetation

The experimental research area is a coastal forest site of a large network spreading the country counting about 42 other forest sites located in the northeast of Tunisia (Figure 1). In addition to Aleppo pine, other species are also planted within this site. During last decades, previous researchers planted provenances, which have been gotten seeds from interior forests. Accompanying species detected here are Kermes oak, Rosemary, Montpelier cists and other species. Suitable forest program in Tunisia is started to discover fitted productive provenances under the same ecological conditions.

Conservation Characteristics of the Experimental Coastal Forest Site

The study zone has an altitude of 180m (a.s.l) and occupy a sub-humid bioclimatic. Latitudinal and longitudinal gradients of the stand are 36°50'N and 8°48'E, respectively. An approximate annual temperature of the study site is of 18°c and the mean rainfall is 540 mm. Likewise, calcareous soil layers with superficial sandy loam substrate were recorded within the site (Table 1).

Experimental Design and Measurements

At the beginning (1963), the species planted zone is estimated to four hectares, where more than thirty rectangular randomized complete blocks were created and planted. Provenances trees are spaced two meter from each other. Prompt, we select a total of 327 tree species representing twelve selected provenances within the study zone for Epidometric measurements *sensu* Ayari et al. (2012a). Afterwards, we determine the average size tree of each provenance. Then, we harvest cone of a limited trees per provenance *sensu* Ayari et al. (2011a). Later and in the laboratory, we calculate the overall species survival rate per planted provenance *sensu* Shupp and Koller (1998). In the next step, we collected mature grey cones to characterise their traits dimensions (Ayari et al. 2012b). Hence, other predicted calculations are done including extracted seeds counting and weighing from each single cone.

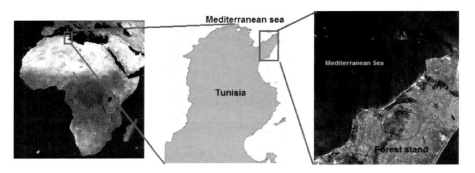

Figure 1. Map showing the geographical location of the coastal forest stand in Tunisia.

Table 1. Overall characteristics of the coastal forest stand and selected provenances indicating their origins (harvested seeds before planting)

Provenances	Prv_1 (H: Sakiet), Prv_2 (J: Jebel Korbus), Prv_3 (O: Oued el Kbir), Prv_4 (R: Berino), Prv_5 (U: Jebel Chehid), Prv_6 (V: Dernaia), Prv_7 (W: Sodaga), Prv_8 (X: Takrouna), Prv_9 (Z: Jebel Koumine), Prv_{10} (AB: Ouesletia South), Prv_{11} (AG: Salloum), Prv_{12} (AH: M'Guilla).
Altitude (m), Latitude (°N), Longitude (°E)	[180; 36°50'; 8°48']
Annual temperature (°C) and rainfall (mm)	[18°, 540]
Soil layers and Bioclimatic zone	[calcareous, Sub-humid]

Statistical Analysis

We have analysed the results using simple regression analyses and descriptive statistics to study possible significant relationships seed-cone productions parameters. During regression analysis, we accepted only regression coefficients describing explanatory parameters to dependant seed and cone production variables with $r > 0.300$ ($p < 0.05$). In addition, we detected differences between provenances using one-way ANOVA and Tukey's Studentized Range test. The programme SAS (2005) was used during statistical analyses.

RESULTS

Aleppo Pine Provenance Endurance and Epidometric Relationships

Overall mean Aleppo pine provenances endurance within Korbus forest experimental site is of about 83%, ranging between 91% and 73% (Figure 2). Subsequently, an overall plantation losses rate across the forest stand is of about 27%. Likewise, provenances U and X coming from Jebel Chehid (district of Beja) and from Takrouna (District of Kef) illustrated within this long period (about sixty years) a high trend for better endurance with lower losses death rate of about 9% (Figure 2). Equally, within this site, the average tree size of selected provenances has an average diameter at breast height (DBH) and an average total height (H_o) measuring $11.6 \pm (0, 3)$ cm and $6.1 \pm (0.1)$ m, respectively. We detect also a high significant relationship between diameter at breast height (DBH) and total height (H_o) if each provenance is treated separately or across selected provenances (Table 2). The recorded correlations coefficients vary between, $r = 0.515$ and $r = 0.867$ respectively for provenance V (Dernaïa Forest (Kasserine district) and provenance W (Sodga Forest (Siliana district)).

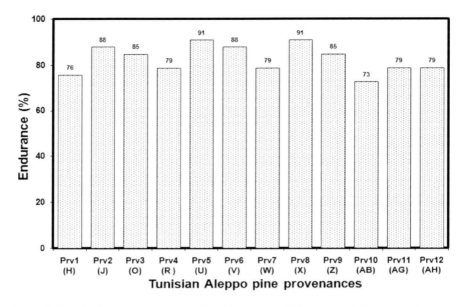

Figure 2. Survival provenances rate after fifty years within a coastal forest stand in Tunisia.

Table 2. Linear significant relationships between total height (H_o) and diameter at breast height (DBH) of selected Aleppo pine provenances within the Tunisian coastal forest stand

Provenance (Prv))	Relationship $H_o = f$ (DBH)	P > Fr
(Prv1 (H))	$Y_H = 0.26\ X_H + 2.67$	0.657***
(Prv2 (J))	$Y_J = 0.17\ X_J + 3.48$	0.530**
(Prv3 (O))	$Y_O = 0.22\ X_O + 3.29$	0.588***
(Prv4 (R))	$Y_R = 0.30\ X_R + 2.32$	0.800***
(Prv5 (U))	$Y_U = 0.31\ X_U + 2.34$	0.709***
(Prv6 (V))	$Y_V = 0.14\ X_V + 3.94$	0.515**
(Prv7 (W))	$Y_W = 0.33\ X_W + 1.73$	0.867***
(Prv8 (X))	$Y_X = 0.27\ X_X + 2.56$	0.725**
(Prv9 (Z))	$Y_Z = 0.21\ X_Z + 3.28$	0.592**
(Prv10 (AB))	$Y_{AB} = 0.22\ X_{AB} + 3.06$	0.697***
(Prv11 (AG))	$Y_{AG} = 0.43\ X_{AG} + 0.85$	0.857***
(Prv12 (AH))	$Y_{AH} = 0.29\ X_{AH} + 2.05$	0.685***
Overall provenances	$Y_{Overall} = 0.26\ X_{Overall} + 2.69$	0.698***

** $p < 0.01$, *** $p < 0.001$.

Tunisian Provenances Productivity Traits

Cones and Seeds Efficiency

For cones and seeds traits within the coastal forest site, we have analysed the results demonstrating an average of tree cone production equal to 51 cones/tree, weighing approximately 922 g/tree (Figure 3a). Nevertheless, if we treated provenances separately, the mean number of produced cone (CNT) varies between height and seventy-nine cones per tree (CWT) weighing respectively about 138 and 2019 g/tree. The utmost production of cones per tree is registered with Aleppo pine provenances coming from M'Guila Forest (the 12th provenance) where the least one is recorded for Sodga Forest (the 7th provenance).

We register intermediate number and weight produced cone per tree for the assessed provenances. Similarly, we observe that minimum produced seed number per tree per provenance is coming from Sodga forest (the provenance: W) with an average mean of about 698 seeds/tree weighing about 7.28 g/tree. Maximum produced seed number per tree is observed with provenances coming from Takrouna Forest where its average produced seeds number per tree is approximately equal to 4808 seeds/tree having an average mass of 82.94 g/tree (Figure 3b). Likewise, we record a produced seed number per cone (SNC) equal to 54 seeds/Cone weighing (SWC) approximately 59 mg/Cone (Figure 3c). Then we identify the highest produced seed number per cone which is AB provenance (the 10th provenance) coming from Forest of South Ouesletia with an average of 65 seeds per cone weighing 62 mg/cone. Minimum seed number produced per cone is observed in provenance coming from Bireno Forest (the 4th provenance: R) with an average of about 37 seeds/cone weighing 40 mg/cone. Maximum produced seed number per cone is observed with provenances coming from Sakiet, Jbel Koumine and Ouesletia south Forests with an average of 65 seeds/cones weighing 59 mg/cone.

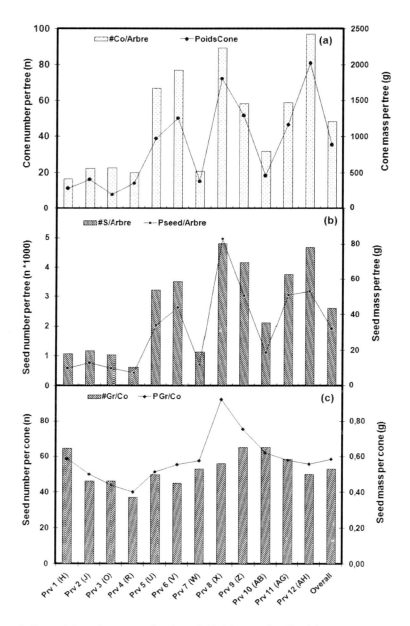

Figure 3. Potential seed cone production of 12 Aleppo pine Tunisian provenances planted before than fifty years ago in a coastal forest stand.

Conservation of Aleppo Pine Forest ... 161

Cones and Seeds Dimension Characteristics

We have analysed the results for cone weight and its dimension, in addition to the average single seed weight (see Table 3). Hence, the average cone size is of about 58.7 mm for its length, 25.1 mm for its width and 17.2 g for its mass. For the produced average seed mass is equal to 11.06 mg. The average cone length (ACL) is ranging between 53.9 mm for Oued el Bir Forests (the third provenances: O) and 67.7 mm registered with provenances coming with Takrouna forests (the eightieth provenances: X). The array of its width is ranging between 23.5 mm and 27.1, registered respectively, with Jebel Chehid forests provenances (the fifth provenances: U) and Takrouna forests (the eightieth provenances: X). The average cone mass (ACM) is also ranging between 11.0 g/cone and 20.6 g/cone registered respectively, with Oued el Bir forests (the third provenances: O) and Jebel Koumine forests (the ninetieth provenances: Z).

Table 3. Mean number and weight of the produced cone per tree registered in coastal Aleppo pine forest stand in Tunisia

Provenances	*ACL* (mm)	*ACWd* (mm)	*ACW* (g)	*ASW* (mg)
Prv1 (H)	61,5 ± (4,9) b[1]	25,1 ± (1,5) bc	16,6 ± (0,9) bc	10.21 ± (2.42) ab
Prv2 (J)	56,3 ± (5,0) c	24,5 ± (1,5) bc	16,9 ± (3,8) bc	10.74 ± (1.38) ab
Prv3 (O)	53,9 ± (0,5) c	24,9 ± (0,1) bc	11,0 ± (3,0) d	12.51 ± (1.16) ab
Prv4 (R)	59,2 ± (2,6) bc	26,5 ± (1,2) ab	20,5 ± (2,1) ab	10.72 ± (1.13) ab
Prv5 (U)	54,3 ± (1,9) c	23,5 ± (0,9) c	15,0 ± (2,2) bc	10.95 ± (0.97) ab
Prv6 (V)	56,4 ± (4,9) c	25,2 ± (1,2) bc	16,1 ± (3,3) abc	9.45 ± (2.33) b
Prv7 (W)	57,7 ± (4,5) bc	24,4 ± (2,0) bc	15,2 ± (3,0) bc	11.13 ± (0.88) ab
Prv8 (X)	67,7 ± (3,4) a	27,1 ± (0,2) a	19,6 ± (1,3) ab	16.70 ± (1.75) a
Prv9 (Z)	58,3 ± (8,2) bc	26,1 ± (1,5) bc	20,6 ± (1,6) ab	11.47 ± (0.50) ab
Prv10 (AB)	56,8 ± (5,6) bc	24,1 ± (2,4) bc	16,8 ± (5,0) abc	10.15 ± (0.68) b
Prv11 (AG)	59,9 ± (2,6) bc	25,9 ± (1,8) bc	18,2 ± (1,5) bc	9.51 ± (0.03) b
Prv12 (AH)	60,5 ± (5,6) bc	25,3 ± (0,9) bc	19,7 ± (2,9) abc	10.68 ± (0.22) ab
Overall mean	58,7 ± (1,2)	25,1 ± (0,4)	17,2 ± (0,8)	11,06 ± (0,46)

[*] Provenance (X: origin), ACL: average cone length, ACWd: average cone width, ACW: average cone weight, ASW: average seed weight, [1]vertical differences between provenances are shown with connecting letters (p < 0.05).

Seed Cone Production Relationships

Analysed results showing the interaction among the utmost significant explanatory variables persuading the Aleppo pine provenances seed-cone production within the experimental site are registered in Figure 4. Overall results showed no significant influence of the diameter at breast height (DBH) and the tree total height (Ho) on the number and weight of conesand seeds (data not shown). However, linear and positive high significant relationships were recorded between cone number per tree and seed number per tree (SNT) (see Figure, 4a), seed weight per tree (SWT) (Figure, 4b) and seed weight per cone (SWC) (Figure, 4c). Recorded correlations coefficient between the aforementioned variables are equal respectively to $r = 0.952$, $r = 0.677$ and $r = 0.427$ (Figure 4). Equally, results showed linear and positive significant ($p < 0.05$) regressions effect of the cone dimension (length (ACL) and width (ACWd)) and average cone weight (ACW) on the produced seed number per tree (SNT), seed weight per tree (SWT) and seed number per cone (SNC). In this case, regression coefficients are balancing between $r = 0.520$ and $r = 0.869$ (Figure 5).

Therefore, the influence of the average cone length on the produced seed number per tree is registered with a positive linear correlation coefficient $r = 0.545$ ($p < 0.01$) (Figure, 5a), where its influence on the produced seed weight per tree and seed number per cone showed correlation coefficients respectively equal to $r = 0.520$ ($p < 0.01$) (Figure, 5d) and $r = 0.713$ ($p < 0.05$) (Figure, 5g). Similarly, the effect of the cone width on the produced seed number per tree recorded a linear positive and significant relationship with an $r = 0.755$ ($p < 0.01$) (Figure 5b) where its effect on the seed weight per tree showed an $r = 0.756$ ($p < 0.01$) (Figure 5e) and with the produced seed number per cone an $r = 0.869$ ($p < 0.05$) (Figure 5h). Likewise, the influence of the average cone weight showed a significant ($p < 0.05$) linear positive effect on the yielded seed number per tree with an $r = 0.633$ (Figure 5c) where its effects on the produced seed weight per tree and seed number per cone showed respectively an $r = 0.705$ ($p < 0.01$) (Figure 5f) and an $r = 0.563$ ($p < 0.05$) (Figure 5i).

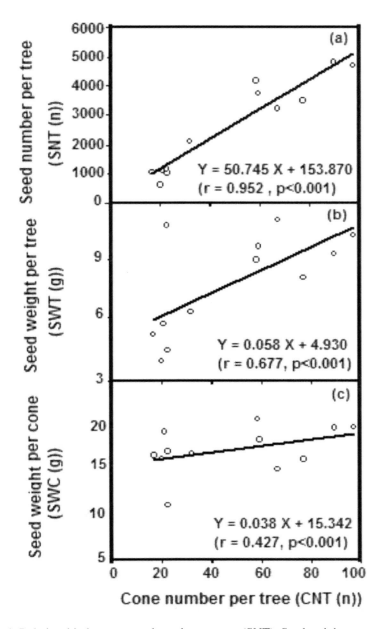

Figure 4. Relationship between seed number per tree (SNT), Seed weight per tree (SWT), seed weight per cone (SWC) and the cone number per tree (CNT) across the selected Aleppo pine Tunisian provenances planted before fifty years ago in a coastal forest stand.

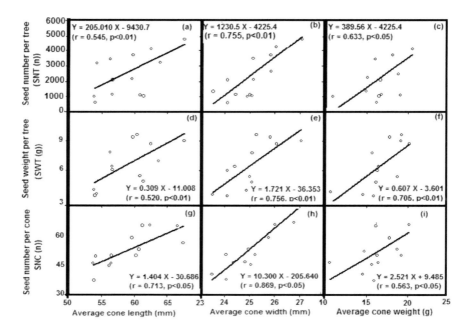

Figure 5. Relationship between seed number per tree (SNT), Seed weight per tree (SWT), seed weight per cone (SWC) and the average cone size (Length (ACL) and width (ACWd)) and cone weight (ACW) across the selected Aleppo pine Tunisian provenances planted before than fifty years ago in a coastal forest stand.

DISCUSSION

Results summarize current research work on the effects of coastal reforestation efforts of several Tunisian Aleppo pine provenances on biodiversity conservation, wildlife habitat management, water resource protection, forest carbon sequestration, sustainable wood, seed and cone production, and reducing the risks and impacts of wildfires. Either in Tunisia or in the Mediterranean basin and other country with localities of similar bioclimatic zone, the Aleppo pine remains the foremost species used in forest protection of poor and eroded soils. This species is also especially used for forest conservation planning of biodiversity and carbon sequestration. Likewise, the Aleppo pine is considered as an important contributor to biotic equilibrium for a large forestry fauna which consuming the disseminated species seeds during the summer period where the temperature is so elevated. Therefore, this species showed previously an endurance ranging between 18.4 and 36.1% according to Weitz (1936). Similarly, according to Bentouati (2006), later Algerian country reforestation efforts using this species have shown endurance ranged between 50 and 80%. Equally, this research work showed that Aleppo pine endurance across selected provenances planted in a coastal forest site could attain 83%. Alike, the endurance within this forest site is balancing between 73% and 91% registered for provenance 10 and provenances 5 and 8 respectively as it is shown in Figure 2. Nonetheless, the high endurance recorded across the assessed Tunisian provenances within the coastal forest site is due to several reasons as explained by Ayari et al. (2016). Even so, the planted Aleppo pine seedlings could avoid any pedo-climatic stress by different mechanisms including deep rooting patterns, strict leaf stomatal controls that avoid substantial water loss, and hydraulic architecture that increases cavitation resistance (Santiago et al. 2004; Nardini et al. 2014). However, the selected provenances showed a good growth rate throughout the last 50 years after the creation of this coastal experimental site to protect forest biodiversity and to conserve the species. During this long period, an overall results illustrated trees with high significant relationships linking total height and diameter at breast height as it is registered in Table 2. Better competitive effect of vital resources (Light, nutrients and water) is mainly observed for both Tunisian provenances (AG and W) coming from Jabel Salloum and Sodgua forest, respectively (Ayari et al. 2016). Previous research work published by Ayari et al. (2011a and 2012b) showed that Aleppo pine trees growing in dense forest site within the Tunisian dorsal mountain chain compete on light, water and nutrients resources availability, which stimulates their growth in height and diameter truck. Furthermore, within this coastal experimental stand, trees competition allocates aerial and rooting growth forms (Ayari 2017).

Coniferous trees phenology, morphology and productivity are mainly controlled by environmental conditions within the growing forest area

(Ayari and Khouja 2014; Ayari et al. 2014). Stand density and basal area may also affect the production potentiality of the pine species (Ayari et al. 2011a). However, bigger crowns Aleppo pine trees are recorded in forests with low density or with isolated trees (Ayari et al. 2012a). This deduction is due to the availability of light and nutrients resources if the density is low which encouraging the trees growing mechanism. Within this coastal stand, the Aleppo pine tree could produce 50 cones/tree weighing 921g/tree. Previously Ayari et al. (2011c) showed that native Aleppo pine forest produced twice cone number (113 cones/tree) than in coastal forest stand. Seed number and weight per tree, in addition to seed weight per cone, are strongly influenced by the number of cone produced by each Aleppo pine tree. Similarly, across the coastal forest stand, the average cone size (length and width) and weight are the most influencing factors either of the number and weight of seeds produced per tree or the number of seeds produced per cone.

Therefore, in this forest coastal stand, the reduced capacity to produce more seed-cone like interior forest of Tunisia may be firstly explained by its high competition between planted provenances for required resources. Secondly, maybe roots of certain provenances have attained rock limestone block causing limited resources nutrients flux. Indeed, critical environmental condition will excite the trees self-control for its vegetative growth than its fruiting (Ayari et al., 2016). Similar concluding clarification is also observed for cone measurements (size and weight) where the Tunisian provenances planted in a coastal forest stand may promote smaller cones production. Current deduction is explained by reduced needles surface area and then reduced photosynthetic rates (Ayari and Khouja, 2014; Ayari et al. 2014). Therefore, vigour of mature harvested cones is influencing extracted seeds quality and quantity. Consequently, within a coastal forest stand, the number of extracted seeds per cone is estimated to the half in comparison to those extracted from cones harvest from native Aleppo pine in Tunisia growing in the dorsal mountain chain. This decrease is linked to the production of small cones size within a coastal forest site where the sea has a negative effect on earlier pollination and later on the cone size measurements.

CONCLUSION

Illustrated results of this research work are very motivating to study the species fruitation within other coastal forested and reforested sites or within the reforested sites of the dorsal mountain chain. Several other concluding remarks will be considered from this study. First, the adaptability of all provenances planted before fifty years ago within the coastal planted and protected forest site. Second, protecting a reforested site elsewhere could increase the conservation of the biodiversity. It's true that the cone seed production within a coastal forest site is more and more reduced in quality and quantity in comparison to a non-coastal forest site, but this production can help to diminish the larger required of the species seeds every year. This work can help to identify important forest areas for in-situ gene conservation and to maintain a tree seed inventory with high quality seed not for Aleppo pine only but for a range of species and then increase production of native plant materials for post-flood or post-fire plantings.

REFERENCES

Antipolis, S. (2001). Tunisia environment and sustainable development issues and policies. Mediterranean country profiles. *Blue plan*, pp. 69.

Ayari, A. (2012). *Effets des facteurs environnementaux sur la fructification de pin d'Alep (Pinus halepensis Mill.) en Tunisie.* [*Effects of environmental factors on the fruiting of Aleppo pine (Pinus halepensis Mill.) in Tunisia.]* Thèse de Doctorat en Sciences Biologiques, Faculté des Sciences de Tunis, 230 p.

Ayari, A., Garchi, S. & Moya, D. (2014). Seed and Cone Production Patterns from Seventy-NineEcotypes of *Pinus halepensis* Mill. Across Tunisia Forests. *Glob J Bot Sci*, *2*, 65, 74.

Ayari, A. & Khouja, M. L. (2014). Ecophysiological variablesinfluencing Aleppo pine seed and cone production: A review. *Tree Phys*, *34*, 426–437.

Ayari, A., Moya, D., Rejeb, M. N., Ben Mansoura, A., Albouchi, A., De Las Heras, J. & Fezzani, Henchi B. (2011a). Geographical variation on cone and seed production of *Pinus halepensis* Mill. natural forests in Tunisia. *J. A. Env*, 403-410.

Ayari, A., Moya, D., Rejeb, M. N., Ben Mansoura, A., Garchi, S., De Las Heras, J. & Henchi, B. (2011b). Alternative sampling methods to estimate structure and reproductive characteristics of Aleppo pine forests in Tunisia. *Forest Sys*, *20*, 348-360.

Ayari, A., Moya, D., Rejeb, M. N., Garchi, S. & De Las Heras, J. (2011c). Fructification and Species Conservation on *Pinus halepensis* Mill. forests in Tunisia: Managing structure Using Individual Tree Size. Chapitre publié dans un livre de Frisiras, C.T: *Pine Forests: Types, Threats, and Management*, Ed. Nova Publishers, USA, pp. 61-80.

Ayari, A., Moya, D. & Zubizarreta, G. A. (2012b). Influence of Environmental Factors on Aleppo pine Forest Production. In Psuatueri and Cannamela *Tunisia: Economic, Political and Social Issues*, Nova Publishers, USA, pp. 93-118.

Ayari, A., Zubizarreta, G. A., Moya, D., Khorchani, A. & Khaldi, A. (2012c). Importance of Non-Wood forest products In Tunisia. In Psuatueri and Cannamela *Tunisia: Economic, Political and Social Issues*, Nova Publishers, USA, 141-153.

Ayari, A., Zubizarreta, G. A., Tomé, M., Tomé, J., Garchi, S. & Henchi, B. (2012a). Stand, tree and crown variables affecting cone crop and seed yield of Aleppo pine forests in differents bioclimatic regions of Tunisia. *Forest Systems*, *21*, 128-140.

Ayari, A., Meftahi, M., Zammeli, F. & Khouja, M. L. (2016). Seed Production variability of Aleppo pine (*Pinus halepensis* Mill.) within Korbus Arboretum (North east of Tunisia). *Glob J Bot Sci*,*4*, 23-23.

Ayari, A. (2017). Aleppo pine seed production of several Tunisian ecotypes planted in a coastal forest stand. *Glob J Bot Sci*, *5*.

BenTouati, A. (2006). *Croissance, productivité et aménagement des forêts de pin d'Alep du massif de Ouled Yagoub (Khenchela-Aures). [Growth, productivity and management of the Aleppo pine forests of the Ouled*

Yagoub Massif (Khenchela-Aures).] Thèse de doctorat d'état en sciences agronomiques. Université El hadj Lakhdar, Batna, *Algérie.*, 116 p.

Chakroun, M. L. (1986). Le pin d'Alep en Tunisie. Options méditerr. *Série Etude CIHEAM*, *86*(1), 25-27.

DGF (Direction Générale des Forêts). (1995). Résultats du premier inventaire forestier national en Tunisie. *Ministère de l'Agriculture*, Tunis, 88 p.

DGF (Direction Générale des Forêts). (2010). Résultats du deuxième inventaire forestier et pastoral national. Inventaire des forêts par télédétection. *Ministère de l'Agriculture*, Tunis, 180 p.

Fady, B., Semerci, H. & Vendramin, G. G. (2003). Euforgen Technical Guidelines for genetic conservation and use for Aleppo pine (*Pinus halepensis*) and Brutia pine (*Pinus brutia*). *International Plant Genetic Resources Institute*, Rome, pp. 6.

Gibbs, J. N., Lipscombe, M. A. & Peace, A. J. (1999). The impact of Phytophthora disease on riparian populations of common alder (Alnus glutinosa) in southern Britain. *Eur J For Path*, *29*, 39-50.

Hall, L. S., Krausman, P. R. & Morrison, M. L. (1997). The habitat concept and a plea for standard terminology. *Wildlife Society Bulletin*, *25*, 173-182.

Hokche, O., Berry, P. E. & Huber, O. (2008). Huber Nuevo Cat. Fl. Vasc. Venezuela. Fundación Instituto Botánico de Venezuela, *Caracas*, pp. 860.

Krugman, S. L. & Jenkinson, J. L. (1974). In Schopmeyer C. S. (tech. coordinator), Seeds of woody plants in the United States. *USDA Agric. Washington, DC. Handbook*. pp. 598-638.

Mezali, M. (2003). Rapport sur le secteur forestier en Algérie. [Report on the forestry sector in Algeria.] *3éme session du forum des nations unis sur les forêts*, pp. 9.

Nardini, A., et al. (2014). The challenge of the Mediterranean climate to plant hydraulics: responses and adaptations. *Environ. Exp. Bot.*, *103*, 68–79.

Naveh, Z. (1974a). Effects of fire in the Mediterranean region. *Fire and Ecosystems* (eds Kozlowsky, T. T. and Ahlgren C. E.). Academic Press, New York, pp. 401-434.

Rigolot, E., Boivin, T., Dreyfus, P., Fernandez, C., Huc, R., Lefevre, F., Pichot, C. & Valette, J. C. (2012). Les pins méditerranéens, Conservation, écologie, restauration et gestion: Défis dans un contexte de changements globaux. [Mediterranean pines, Conservation, ecology, restoration and management: Challenges in a context of global change.] Synthèse des travaux de MedPine 4, IVe Conférence internationale sur les pins méditerranéens. Avignon 6-10 Juin, 2011. *For méditerr, 31*(1), 3-18.

S.A.S. (Statistical Analysis Systems). (2005*). SASR User's Guide* v 9.0. Statistical Analysis Systems Institute. Inc., Cary, North Carolina.

Santiago, L. S. et al. (2004). Coordinated changes in photosynthesis, water relations and leaf nutritional traits of canopy trees along a precipitation gradient in lowland tropical forest. *Oecologia, 139*, 495–502.

Schupp, J. R. & Koller, S. I. (1998). Growth and productivity of disease-resistant apple cultivars on M.27 EMLA, M.26 EMLA and Mark rootstocks. *Fruit Var J, 52*, 150-154.

Spencer, D. (2001). Conifers in the dry country. *A report for the RIRDC/L and W, Australia/FWPRDC.* Joint venture agroforestry program, RIRDC publication N°01/46, RIRDC project N° CSF-57A, pp. 60.

Weitz, J. (1936). Le pin d'Alep en Palestine [The Aleppo pine in Palestine], *Silvae mediterranea*, pp. 110-141.

In: Forest Conservation
Editor: Pedro V. Eisenlohr

ISBN: 978-1-53614-559-5
© 2019 Nova Science Publishers, Inc.

Chapter 6

LINKING AGROFORESTRY TO REDD+ ACTIVITIES IN THE AMAZON: IMPLICATIONS FOR BIODIVERSITY AND CARBON CONSERVATION

Pedro Manuel Villa[1,], Sebastião Venâncio Martins[2],*
Silvio Nolasco de Oliveira Neto[2],
Alice Cristina Rodrigues[1]
and Lucieta Guerreiro Martorano[3]

[1]Department of Botany, Federal University of Viçosa,
Viçosa, MG, Brazil
[2]Department of Forest Engineering,
Federal University of Viçosa, Viçosa, MG, Brazil
[3]EMBRAPA Eastern Amazon, Santarém, PA, Brazil

---

[*] Corresponding Author Email: pedro.villa@ufv.br.

INTRODUCTION

The Amazon basin is the most important extension of continuous tropical forest in the world, including nearly 40% of the total area of tropical rainforests distributed among nine countries in South America (Laurance et al. 2001; Cerri et al. 2007), and has about 11% of the tree species estimated to occur worldwide (Cardoso et al. 2017). Its impact at a global level has been attributed to its role in the maintenance of biodiversity and hydrological cycles, as well as to its role in terrestrial carbon storage (Malhi et al. 2006; ter Steege et al. 2013; Fauset et al. 2015). Currently, as a result of widespread preoccupation associated with the future of this region, investigations of ecosystem patterns and processes in these forests are helping to understand the effects of ongoing land use changes, which are causing atmospheric emissions of carbon and biodiversity loss (Brienen et al. 2015; Barlow et al. 2016, Nobre et al. 2016). Furthermore, there are different possible consequences, including forest degradation and the decline in the quality of life of local populations, both of which have been poorly examined so far.

Compared to other tropical forests in different latitudes in the world, the Amazon is typically known as a biogeographical region having a high biological diversity (Strassburg et al. 2010). The Amazon forest is a place where numerous species of plants have been domesticated for diverse purposes, such as a source of food (Levis et al. 2017), and it is a place where crop-growing practices have existed among indigenous populations for many hundreds of years or more (Piperno 2009; Arroyo-Kalin 2012; Bush et al. 2015). Likewise, the Amazon is potentially an efficient source of carbon sequestration, providing the world with an environmental service (Betts et al., 2008; Fauset et al. 2015), by storing up to 90 Pg of carbon, both below- and aboveground (Malhi et al. 2006; Betts et al. 2008). In this sense, and since Brazil represents the country with the largest extension of area of the Amazon with more than 60% of the whole basin, it is a strategic country for biodiversity conservation purposes and for carbon storage (FAO 2010). However, there is still little information on the patterns related to the production of ecosystem goods and services at a regional level, and little information on their impact on the quality of life of the surrounding indigenous and non-indigenous communities.

There is evidence on how the continuous changes in land use during recent years are producing transformations in the structure and functioning of Amazonian ecosystems (Laurance 2004; Lewis et al. 2015; Nobre et al. 2016), which, at the same time, has a considerable negative impact on human settlements (Boucher et al. 2013; McMichael et al. 2017). Accordingly, continuous growth of deforestation of Amazon forests by means of the establishment of agricultural systems in indigenous and non-indigenous communities is also causing greater pressure on the forests, given the enlargement of the agricultural frontier and the increase of production yields that serve local, as well as national and international markets (Boucher et al. 2013; Rodrigues-Filho et al. 2015; Villa et al. 2017). Besides the areas dedicated to livestock farming and extensive soy crops (Boucher et al. 2013; Nepstad et al. 2014), among the types of agricultural practices in the region, shifting cultivation (slash and burn) is prevalent and consists of small areas of deforestation of primary and secondary forests, as part of the necessary land preparation for the establishment of annual crops, mostly for short periods of time (2-3 years), due to diminished soil fertility (Oliveira et al. 2011; Arroyo-Kalin 2012). For this reason, the conservation and sustainable management of forests in the Amazon has become important during the last decade, as ways to prevent and reduce the effects of accelerated deforestation and forest degradation caused by the different changes in the ways of using land throughout the ecological history of the basin (Nepstad et al. 2009; Aragão et al. 2014; Nobre et al. 2016; McMichael et al. 2017).

The future of the conservation of the Amazon basin heavily depends on the strategies and actions created to optimize the use and sustainable management of ecosystem goods and services, with agriculture being one of the main direct causes for deforestation; and such strategies ought to guarantee a simultaneous optimization of environmental, social and economic impacts (Villa et al. 2015, 2017). This is why, by searching for a balance between development and environmental conservation, agroforestry has entered into proposals for sustainable agricultural models for the region (Porro et al. 2012; Villa et al. 2017). This land use type has also recently been recognized as a model of Intelligent Climate Agriculture that could contribute to the increase of an ecosystem's productivity, so that means of livelihood are produced for the current and future generations, and that simultaneously greenhouse gas emissions are reduced (FAO 2013). However, there is no proper analysis fully reconciling agroforestry with biodiversity conservation and carbon sinks, as fundamental goods and services for the well-being of local populations, and particularly, as an alternative to reduce the pressure caused by the use and management of the forests.

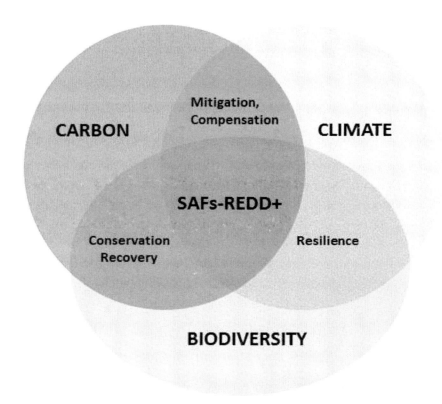

Figure 1. Relationship of the impacts generated by agroforestry systems on conservation, restoration and management of biodiversity and carbon in Amazonian forests, and its relation to climate as a strategy to maintain a functional balance in the production and conservation of goods and services ecosystem. Adapted from Kapos et al. (2012).

With this review, we aimed to develop an analysis of the potential environmental impacts of agroforestry as a REDD + alternative, through reduction of emissions from deforestation (i) and forest degradation (ii), enhancement of carbon stocks (iii), sustainable management of forests (iv), conservation of carbon stocks (v), and through the integrated implementation of REDD+ strategies in the Amazon basin (Figure 1). To date, there is only one extensive review on the impact caused by the management of the forests and the land, both on biodiversity and carbon, where the applicability of the proposed measures to reduce the emissions due to deforestation and forest degradation was examined. However, there is no comprehensive proposal which would simultaneously analyze the potential elements and the existing relations between agroforestry as a sustainable alternative to manage the Amazon forests, and biodiversity and carbon as important sources of environmental goods and services, given the continuous changes in the use of land, caused by deforestation practices carried out with the purpose of establishing shifting farming systems, which at the same time, bring as a consequence the fragmentation and the degradation of the diverse landscapes of the Amazon region. To address the need to provide alternatives to increase the efficiency of the initiatives of management and conservation of biodiversity and carbon in vulnerable areas thorough REDD+ activities, the objective of this review is to develop a comprehensive analysis of the potential social and environmental impacts that agroforestry could bring to the Amazon basin.

LAND USE PATTERNS IN THE AMAZON BASIN

The Amazon region holds a wide variety of ethnic groups that have regarded their territories as the place where diverse social and cultural relations occur, according to the natural resources of each settlement (Roosevelt 2013; Bush et al. 2015; Piperno et al. 2015). Subsequently, given the broad cultural indigenous spectrum and the different types of forests existing within the basin, multiple types of use and management of ecosystem goods and services provided by these forests have been recognized (Levis et al. 2017); however, this has barely been studied so far. Nevertheless, the gathering of non-timber forestry products and the food production carried out with shifting cultivation systems, as main traditional local means of livelihood, are the general pattern at a local level (Arroyo-Kalin 2012; Roosevelt 2013; Levis et al. 2017). These ways of using the land have been predominant in tropical forests at a global level, particularly when used by indigenous societies that have developed strategies to adapt to their surrounding environmental conditions (Barton et al. 2012).

According to archeological evidence, tropical rain forests have been exposed to a constant pressure coming from deforestation processes and local populations' subsistence practices, during long periods of time (Barton et al. 2012; Piperno et al. 2015); conversely, during the last decades, forest loss rates have significantly increased, as a result of shifting cultivation, livestock farming and timber extraction practices (Lewis et al. 2015; Andela et al. 2017). Similarly, it has been well documented that the Amazon occupation began back in the Holocene, following ancient evidence of the use of fire by means of establishing agriculture systems (carbon from burning of vegetation (charcoal) and pollen in soil samples), even if it has been difficult to understand its complexity (Arroyo-Kalin 2012; Piperno et al. 2015; Goulart et al. 2017). In this respect, various authors have undertaken examinations through which the importance of shifting cultivation was explained and that have enhanced recent discussions on Pre-Columbian Amazonia, specifically by using findings from Pre-Columbian anthropogenic soils that have helped re-evaluate how the speed and heterogeneity of regional crop domestication (specific growing practices) might have influenced the history of fire in the Amazon (Arroyo-Kalin 2012; Goulart et al. 2017). Therefore, in this context, understanding of the effects of the historical use of the land and its effects on possible future scenarios for Amazonian forests is a complex subject.

Only since the last century, after 1960, a complex process of transformation of the agricultural, environmental, social and cultural contexts of the Amazon basin began to be carried out, while there were changes in agricultural policies through a migration program in some countries of the basin (Llambi and Llambi 2000; Diniz et al. 2013;

Rodrigues-Filho et al. 2015). Agricultural reforms in these countries were oriented toward the widening of the agricultural frontier, in order to increase the levels of production, and satisfy national demands for agricultural products, and the occupation of new territory was also stimulated by geopolitical strategies (Diniz et al. 2013; Villa et al. 2017). In this context, recent findings show that agrarian reform policies contributed directly to carbon loss. Therefore, the establishment of new settlements implies possible losses of stored carbon, especially if settlements are created in areas with high carbon levels (Yanai et al. 2017). However, there is little evidence on how these accelerated processes of demographical expansion and historical occupation of the diverse indigenous communities in the basin took place, as well as there is little evidence on the subsequent processes of immigration of the non-indigenous populations and emigration of indigenous populations towards other communities.

Also, the first impacts caused by the influence of Western civilization on the patterns of land and forestry resource use have not been taken into account. Historical patterns of occupation of the indigenous and creole populations are thought to have had a different history in each eco-region, given the current political and economic development processes in each country that is part of the basin. For instance, in some of the states in the Brazilian Amazon, evidence has been found for occupation changes through different models for deforestation, which reveals important changes in the patterns of human occupation (Espindola et al. 2012; Rodrigues-Filho et al. 2015; Aguiar et al. 2016). Nevertheless, regardless of the historical processes of occupation, and of the different development processes that have occurred at a regional scale, there is currently a prevailing tendency towards settlement of indigenous populations in specific communities, besides a growth and expansion of those communities towards other regional settlements (Rodrigues-Filho et al. 2015; Villa et al. 2017). In this way, such social and demographic processes have also brought as a consequence a negative impact with respect to the ecosystem goods and services that these local settlements demand, maintaining a constant rate of deforestation as a result of the establishment of shifting agriculture systems that have been increased at the same speed as local settlements. Also, because of the low fertility of the soil, the whole situation has forced the locals to settle in new communities (Arroyo-Kalin 2012; Bush et al. 2015; Villa et al. 2017). Furthermore, the exact strength of the existing relation between forest sustainability and nomadic indigenous populations' emigration, or between a population's growth and an agriculture frontier's expansion is not well known.

In this way, several researchers point out that the structure and dynamics of the ecosystems and human settlements have a significant effect on the quality of life of the local populations. This could be evaluated through the social and environmental quality of life indicators, so that those tendencies and effects that have been produced by the changes in the use of land, mainly because of deforestation, could be better understood. Such social and ecological processes have produced a greater dependency on food sources, both local (traditional slash and burn systems), and from outside (the influence of Western civilization), which altogether, have simultaneously been leading to a massive change in life strategies of the ethnic groups.

For this reason, recently some researchers have suggested that to increase the efficiency in the application of REDD+ strategies in Amazonian forests, it is fundamental to simultaneously understand the relationship of patterns and processes of forest ecosystems within the socio-ecological system, through the analysis of biophysical and anthropogenic predictors (Figure 2), from local to regional scales (Villa et al. 2017). Likewise, the simultaneous evaluation of the anthropogenic predictors that directly affect the current and future demands for ecosystem goods and services has been proposed, mainly under dynamic migration, colonization and population growth scenarios. These predictors of the evaluation of REDD+ actions represent current critical points to increase the efficiency of conservation and management of forest ecosystems in the Amazon. For this reason, we must begin to evaluate the magnitude of the effects that predictors of deforestation have, especially those related to agriculture, for the integration of REDD+ actions through the rehabilitation of degraded areas with successional and permanent agroforestry systems, as well as with the application of different restoration methods (passive and active) and integrated management of forest resources (*sensu* Villa et al. 2017).

Nonetheless, up to now, little has been discussed or analyzed regarding the potential role of agroforestry systems (SAFs) in such different scenarios of land use in the Amazon. Agroforestry is considered as an alternative that could eventually reduce the pressure exerted by deforestation of new areas with agricultural purposes, through the rehabilitation of degraded forest areas, enhancing previously established productive fallow land management, and facilitating a passive restoration during the successive natural regeneration of the secondary forest. This would be a simultaneous contribution to the conservation and recovery of biodiversity and carbon sinks, meaning a greater production of goods and services that would help to create a better quality of life for the local populations.

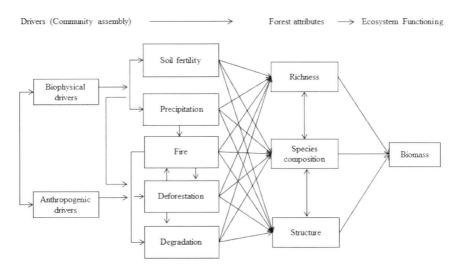

Figure 2. Conceptual diagram indicating how biophysical and anthropogenic predictors affect the structure of plant communities, and how the predictors and plant communities' **structure** simultaneously affect carbon storage in biomass. Adapted from Poorter et al. (2017) and Villa et al. (2017).

SHIFTING CULTIVATION IN THE AMAZON

Shifting cultivation, or slash and burn practices, has been ancestrally performed by many ethnic groups in the Amazon basin (Arroyo-Kalin et al. 2012; Bush et al. 2015), and it has consisted in the deforestation of small areas of primary and secondary forests (<1 ha) with the purpose of establishing traditional plantations of bitter manioc root (*Manihot esculenta* Crantz) as the predominant crop (Jakovac et al. 2016.). Thus, shifting cultivation consists of a unit of production for domestic subsistence purposes in different indigenous settlements. Besides the food gathering practices in the forests, and the use of hunting as a source of protein, the shifting agriculture systems constitute the main means of production of ecosystem goods and services to partially meet the needs of families among the indigenous and non-indigenous communities in the Amazon basin, taking advantage of the fertile soil (at the initial stage of cultivation) and the local agroforestry biodiversity. Such agriculture systems may be called differently with indigenous or creole words, according to the specific region, and these have been broadly recognized as "chakras" in Ecuador, "conucos" in Venezuela, and also, as "coivaras" in Brazil (Villa et al. 2015).

LINKING REDD+ACTIVITIES: METHODS AND CHALLENGES

Negative scenarios and drivers that threaten Amazon forests must be reversed through comprehensive, rapid and feasible actions to reduce emissions from deforestation and forest degradation, as well as comprehensive activities for conservation, sustainable management and increased carbon reserves (REDD +) within the framework of the agreements of the United Nations Convention on Climate Change (Jagger et al. 2013). Likewise, these processes have important implications in the REDD+ context whose main objective is to promote low emission development trajectories, increasing the value of forests in relation to different types of land use (Angelsen and McNeill 2013). For this reason, it is emphasized that maintaining the REDD+ objectives requires the transformation of economic activities inside and outside forests, which are recognized as the direct and indirect causes of deforestation and forest degradation (Angelsen and McNeill 2013; Jagger et al. 2013).

Recently, Villa et al. (2017) concluded that although there are alternatives to reduce emissions from deforestation and forest degradation, there is still little research on methods for assessing and monitoring the relationship between these processes and biodiversity (REDD+), as well as the impacts generated by the socioecological system on the stability and recovery of ecosystem services. With this review, we aimed to (i) describe the relationship between biodiversity and ecosystem functioning; (ii) analyze the possible synergistic impacts of biophysical and anthropogenic predictors of deforestation; and, in view of that, (iii) propose the integrated implementation of REDD+ strategies in the Amazon basin.

Two main strategies are recognized to mitigate the increase of carbon in the atmosphere due to deforestation: i) controlling the reduction of emissions through the conservation and sustainable management of forest ecosystems, and ii) increasing production of plant biomass as an important carbon reserve in the biosphere, through forest restoration practices and rehabilitation of degraded forest areas as part of the comprehensive REDD+ activities (Kapos et al. 2012; Villa et al. 2015). Implicit in these strategies is a positive impact on biodiversity, although it would be more convenient to emphasize the importance of the biodiversity-ecosystem function balance in terms of REDD+.

The proposal to link REDD+ activities arises in the face of growing threats of biodiversity loss and carbon stocks due to changes in land use and climate change. For example, it has recently been determined that Amazon forests, after having been important carbon sinks for the past 30 years, could begin to lose their potential capacity for long-term carbon accumulation, and then gradually become carbon sources to the atmosphere due to the increase in tree growth and mortality rates, which could lead to a shortening of carbon residence times (Brienen et al. 2015).

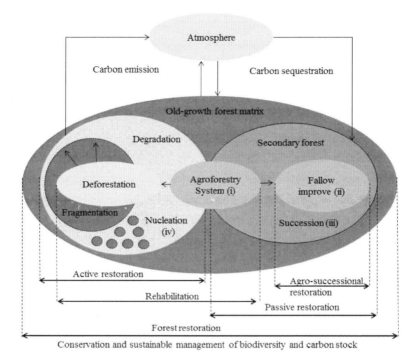

Figure 3. Integration of different types of forest management with impacts on REDD+ actions such as the reduction of carbon emissions from deforestation and from forest degradation, through conservation of old forests, sustainable management of second growth forests, and increase in carbon stocks and biodiversity. The following rehabilitation and restoration alternatives are proposed: agroforestry systems as a rehabilitation method (i), improved fallow as an agrosuccessional system (ii), succession as a passive restoration mechanism (iii), and nucleation as active restoration (iv), and old growth and second growth forest conservation (v). Villa et al. (2017).

This evidence indicates that if mature forests do not guarantee stable carbon storage and agriculture continues to be one of the main causes of carbon emissions, REDD+ activities should be implemented that go beyond forest conservation and emission reduction; that is, activities that also contribute to the increase of carbon stocks. In this sense, it is expected that agroforestry systems and regeneration of secondary forests can benefit from carbon enrichment in the atmosphere to increase their growth and productivity. For this reason, four urgent REDD+ activities are also proposed to recover and increase carbon stocks and biodiversity in the Amazon within a forest management system (Figure 3). Improved fallows through controlled succession are considered after a cycle of shifting cultivation as a measure of agrossuccesional restoration.

Likewise, forest restoration in the Amazon is proposed as part of the REDD+ strategies and must transcend the manipulation of the structure of ecological communities according to their functional potential in the ecosystem. This criterion is also applicable to agroforestry systems where it is possible to select species according to the variability in their functional attributes (tree size, carbon storage capacity, nitrogen fixation, decomposition and nutrient cycling, and food provision for wildlife), which could play a fundamental role in ecosystem functioning (Villa et al. 2017), in addition to supporting the production of food as a sustainable way of life for local people. This type of approach is quite limited for SAFs; however, we suggest that the way of managing the structure of SAFs, including spatial planning of productive landscapes with agroforestry systems, could be an effective approach to improve productivity in these systems, which in turn will depend on the integrated management of forest resources.

Finally, it is important to note that carbon in protected areas has a strategic value for environmental conservation and climate change mitigation because these areas have less emission risk than carbon stored in vegetation located outside protected areas, although the effectiveness of protected areas varies according to the demographic pressure of each sub-region of the Amazon basin (Nogueira et al. 2017).

REDUCTION OF EMISSIONS FROM DEFORESTATION AND DEGRADATION

The intense and accelerated changes in land-use in forested areas in the Amazon are causing serious negative consequences in the structure and functioning of ecosystems, principally due to forest degradation and deforestation due to the expansion of the agricultural frontier (Nobre et al. 2016; Yanai et al. 2017). Among the consequences are biodiversity loss (Barlow et al. 2016; Solar et al. 2016), as well as a reduction in the capacity to sequester and store carbon which could lead to an increase in the emission of greenhouse gases (Silveiro et al. 2015; Anderson-Teixeira et al. 2016), causing alterations in hydrological mechanisms and the local, regional, and global climatic regimes (Devaraju et al. 2015; Lawrence and Vandecar 2015). Furthermore, the changes could provoke a taxonomic and functional simplification of these forested ecosystems due to the convergence of species in the landscapes of the region. This process is called biotic homogenization (Olden and Roone, 2006; Baiser et al. 2012), and is little studied in tropical forests under intense anthropogenic pressure (Arroyo-Rodríguez et al. 2013; Solar et al. 2015).

In this sense, successive cycles of shifting cultivation (two or more cycles in the same area), which is still the standard land-use type in the Amazon, can induce floristic homogenization due to forest degradation. This process occurs due to the loss of regeneration capacity caused by frequent cycles of fire that each year reduces species diversity that exists in the natural seed bank in the soil (Figure 4). These losses due to deforestation of primary or secondary forests successively reduce the resilience of the soil, and ecological processes are disturbed, coupled with dissipation and degradation of the regenerative processes of the forest cover due to reduction of the source of seed propagation (Gandolfi et al. 2007). Furthermore, these processes are exacerbated by intense losses of superficial soil layers, principally in areas with steep slopes that are highly exposed to erosive processes (Martorano et al. 2017).

The response of the terrestrial ecosystem to the increase of atmospheric CO_2 levels and to climate change provides an important feedback mechanism for future global warming (Booth et al, 2012). For example, one of the planet´s most important carbon sinks is the Amazon forest which stores about 150-200 Pg C in live biomass and soils (Saatchi et al. 2011; Feldpausch et al. 2012). The significant increase in carbon in the atmosphere has caused important changes in the accumulation and dynamics of carbon on a global scale (Huntingford et al. 2013). Amazonian forests and their teleconnections (Brienen et al. 2015) with other regions have simultaneously been demonstrating evidence of their effects on climate changes at an alarming rate (Devaraju et al. 2015; Lawrence and Vandecar 2015).

Nonetheless, with the two proposed strategies to mitigate the increase of carbon in the atmosphere, i) control emission reduction through management of sources, and ii) strengthen important carbon sinks by increasing vegetation biomass (IPCC 2000), it is also possible to advance towards the consolidation of REDD+ policies, especially with the adoption of agroforestry systems as alternatives to the traditional system of slash and burn, thus promoting a redesign of the landscape augmenting vegetation corridors and ecological connectivity, and most importantly, reducing the deforestation and degradation of forests in the region.

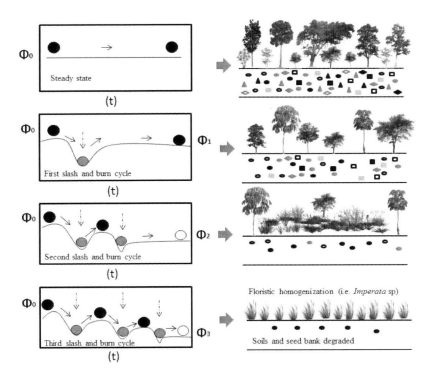

Figure 4. Classical model representations of ecosystem resilience. Balls represent states of an ecosystem (where a state is defined by a set of variables such as biomass and species richness), continuous arrows represent disturbances (clear-cutting for shifting cultivation), valleys represent biomass or richness loss, and peaks represent thresholds (biomass or richness recovery, dashed arrows). Villa et al. (2017).

RECOVERY AND INCREASE OF BIODIVERSITY AND CARBON STOCKS

With respect to carbon recovery in a forest landscape matrix, where there are areas with different levels of conservation and management, it is possible to implement different REDD+ activities. Consequently, according to the potential of agroforestry, three strategies have been proposed for recovering and increasing carbon stocks in Amazonian forests (Villa et al. 2015): i) rehabilitation of areas that have been degraded due to successive cycles of fire use in the same production unit for shifting cultivation purposes and with little fallow time between cycles (Figure 5); ii) passive restoration is indirectly facilitated by avoiding the successive use of forests with different succession stages after shifting cultivation cycles; and iii) sustainable management of secondary forests could also be implemented through improved fallows, increasing the planting density of long-cycle agroforestry tree species (*sensu* Villa et al. 2015).

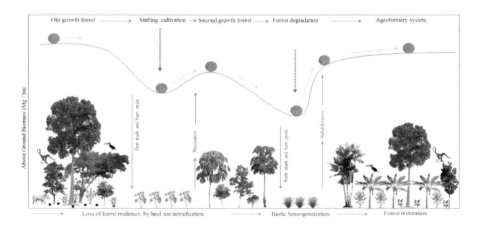

Figure 5. Potential accumulation of biomass and biodiversity through a land use intensification gradient and forest management with impacts on REDD+ activities, mainly in multistrata agroforestry systems dominated by fruit trees, timber trees, and biennial species.

Linking Agroforestry to REDD+ Activites in the Amazon 187

Presently, agroforestry systems in protected areas in Brazil represent an important alternative involving environmental, social and economic aspects, to help to create and maintain sustainable management of the forests in small rural indigenous properties (Soares-Filho et al. 2010; Oliveira-Neto 2012; Porro et al. 2012), which would represent a point of reference and an alternative as an administration model for other countries that are part of the Amazon basin, even if certain types of use of land are recognized to be producing significant impacts on forest loss (soy, livestock, mining, selective logging). Hence, agroforestry as an alternative to the model of shifting cultivation may represent a suitable option for the conservation and management of primary forests, along with the different stages of succession and regeneration (secondary forests), which are important for the recovery and maintenance or increase of the existing biodiversity and of the carbon sinks, as a means to provide the region and the globe with an important service (Nair et al. 2009, 2010; Sougata and Shibu 2012; Villa et al. 2017). Furthermore, agroforestry may help small indigenous and non- indigenous farmers in the Amazon region, by providing them with an alternative to harvest agroforestry products and contribute to their own food security, both at domestic and local levels, as an alternative to create income by means of placing their products on the market in local agro-food networks, or in organized national networks.

AGRO-SUCCESSIONAL RESTORATION

In recent years research on the development of Analogous and Regenerative Agroforestry Systems (ARAFS) has increased in the Amazon region, especially with respect to the synchronization and spatial arrangement of species, but it is still necessary to investigate the relation between ARAFS and time and to identify strategies capable of increasing the spatial scale of the arrangements in integrated production systems (Martorano et al. 2016), in order to promote the adoption of these systems among agricultural producers. Among the possibilities presented by these systems is the establishment of different species that could simulate the same dynamic as the sequence of phases of natural forest regeneration, but in this model there is a purposeful time sequence of species based on their life cycles (Kato et al. 2006; Vieira et al. 2009). In many regions in the Amazon, these models could be initiated starting from a natural fallow, thus taking advantage of the productivity of annual crops associated with shrubs that have a short life cycle, and then the followed by long-lived species such as fruit trees and timber trees, thus systematically establishing native species forest cover (Villa et al. 2015).

These agroforestry arrangements can be highly productive, many times more than unproductive fallows, with a large accumulation of litter on the soil surface and a high level of carbon storage in the above-ground biomass. However, there are places such as traditional communities in the State of Amazonas where fallow systems incorporate native species that grow simultaneously with traditional agricultural crops, but without a systematic spatial arrangement that could, in theory, improve the use of the same area, in terms of fertility and production. For this reason the current work proposes the use of spatial management in order to increase the density and diversity of agroforestry species, also known as "enriched fallow" through "species enrichment" (Villa et al. 2017). This proposed agroforestry arrangement of species would tend to increase the production and income compared to a monoculture, degraded secondary forests, and even forest plantations (Figure 5).

Furthermore, an important concept was recently proposed (Villa et al. 2015, 2017), which has some basic premises that deal with the establishment of agro-successional systems, and alternatively, agroforestry systems in degraded areas. These models could have a substantial effect on the mitigation of environmental impacts to the degree that these systems become permanent elements of an ecosystem. In this case, the basic premise is the planned management of successive cycles of different components of an agroforest, with respect to its phenological and productive phases, through the subdivision of areas with different development chronosequences within the same agroforest, but this would require agrosilvicultural planning and management. Using this method it would be possible to, in the future, systematically plan and substitute each senescent individual from perennial species for individuals that are in an initial or intermediate development phase. The objective of this system would be to maintain the productivity of annual species that are cultivated together, for each management cycle in each sub-divided area, up to the complete closing of the agroforest canopy (Villa et al. 2017).

However, active restoration also could be simultaneously implemented in degraded forest areas within the same landscape unit, through nucleation with the transposition of the seed bank from nearby reminiscent forests. These actions could be included in REDD+ in temporary agroforests or in fallow fields that have experienced a decrease in productive capacity, and can increase the accumulation of soil organic matter to create buffer areas within the larger forest matrix, or even connective corridors between forest fragments and the larger forest matrix (Figure 3; Villa et al. 2017).

PASSIVE RESTORATION

Amazon old growth forests and their biodiversity are seriously threatened due to land-use change in the Amazon, because agricultural expansion and intensification by shifting cultivation is one of the main causes (Barlow et al. 2016; Jakovac et al. 2016). Accordingly, constant growth of deforestation of Amazon forests by shifting cultivation in indigenous and non-indigenous communities is also causing a greater pressure on the forests' recovery, given the increase of the production yield serving domestic demands, as well as local, and even, national markets (Jakovac et al. 2016; Villa et al. 2017). Despite this, secondary forests that are regrowing after abandonment of farming systems (i.e., shifting cultivation) still represent a reservoir of biodiversity (Gibson et al. 2011; Chazdon 2014) and biomass stock (Chazdon et al. 2016; Poorter et al. 2016), since these forests contain more than half of the global forest area (FAO, 2010). Likewise, the importance of secondary succession has already been demonstrated as a feasible method for forest restoration with greater floristic diversity, when compared to other methods, where a limited number of tree species can be used (Engel and Parrota 2001; Palomeque et al. 2017).

The process of diversity recovery during secondary succession depends on several environmental drivers that determine its trajectory (Arroyo-Rodríguez et al. 2015; Derriore et al. 2016; Meiners et al. 2016). However, in human-modified landscapes, the land use history (Klanderud et al. 2010), intensity and duration of land use (Guariguata and Ostertag 2001), as well as the time since abandonment (Pascarella et al. 2000) are some of the most important anthropogenic drivers for tropical forests (Chazdon 2014). Thus, there is some evidence that the recovery of biodiversity could be high because the species richness can quickly recover to mature forest levels during succession (Norden et al. 2009; Martin et al. 2013; Villa et al. 2018a). On the other hand, the recovery of species composition could take centuries according to the environmental conditions of each locality (Finegan 1996; Dent et al. 2013).

In this context, it has been well demonstrated that there is a significant effect of land-use dynamics on the biomass and carbon recovery capacity of secondary forests (Chazdon et al. 2016; Poorter et al. 2016). This ability to recover biomass depends mainly on climatic conditions and disturbance types (Poorter et al. 2017; Villa et al. 2017). It has recently been shown that precipitation and edaphic gradients are important drivers of ecosystem functioning in neotropical forests (biomass stock and dynamics), where water availability has a positive effect on the production and storage of biomass, and a future expected increase in atmospheric drought could, therefore, potentially reduce carbon storage (Poorter et al. 2017). Likewise, it has been demonstrated that these climatic processes can alter the gradient of biomass production in the Amazon basin (Mitchart et al. 2014).

On the other hand, it has also been evident that the intensification of land use considerably affects the capacity to recover biodiversity and biomass in forests of the Amazon (Jakovac et al. 2015, 2016; Villa et al. 2018b). In these studies, the authors determined how the recovery of the forest structure was affected by the intensity of management, while the recovery of the diversity of species was driven by the configuration of the landscape. In this regard, it has been demonstrated that short shifting cultivation cycles can result in progressive loss of nutrients and biomass recovery capacity (Oliveira et al. 2011; Jakovac et al. 2016). For example, after the first cycle of land use, rates of carbon accumulation in above ground biomass tend to be faster during the first two decades of succession of Amazonian forests (Poorter et al. 2016). However, after two or more cycles of shifting cultivation secondary forests gradually lose their ability to recover biodiversity and biomass, in some cases reaching extreme limits of degradation (Chazdon 2014; Villa et al. 2017).

The net primary productivity in secondary forests can be three to five times higher than that of primary forests due to the high demand for carbon sequestered for biomass accumulation; however, the carbon reserves in secondary forests are lower than in primary forests (Luyssaert et al. 2008; Poorter et al. 2016). Globally, secondary forests are an important carbon sink that partially compensates for carbon emissions caused by tropical deforestation (Houghton et al. 2000; Feldpausch et al. 2004).

All these patterns and processes related to the secondary succession of forests are important starting points to consider passive restoration as a REDD+ alternative, as well as to improve criteria for forest management for the purpose of active restoration. For this reason, making a scale analysis of the Amazon basin, it is expected that the capacity of forest recovery depends directly on the climatic gradients (precipitation and soils), and intensification of land-use gradients (Figure 6). Considering this pattern as a basic principle for REDD+ activities, it is expected that the humid forests of the Amazon (i.e., rainforests in the northwest region of the basin) with low intensity of use, will have greater capacity to recover biodiversity and biomass, compared to dry forests that experience greater pressure by anthropogenic drivers (i.e., seasonal forests in Southeast region of the basin). To improve the efficiency of REDD + activities, both passive and active restoration in Amazonian forests, it is essential to consider this existing balance between gradients of disturbances and climatic gradients, which will also determine a gradient of resilience (Figure 6).

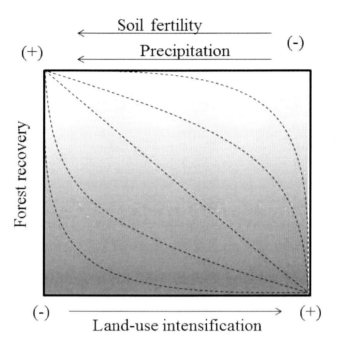

Figure 6. Conceptual diagram indicating theoretical trends of biophysical predictors (precipitation and soil fertility) and land use intensification inducing biomass loss and forest degradation.

IMPROVED FALLOW

The fallow system consists in letting the soil rest for some time, according to the ancestral indigenous patterns, starting at the moment the agricultural systems stop being managed. Afterwards, they are abandoned because of the gradual loss of natural fertility of the soil, coupled with a reduction in the productivity of the annual crops (Huber and Zent 1995; Villa et al. 2012; Jakovac et al. 2015; Villa et al. 2018). In some cases, the fallow system starts after the second or third cycle of planting-harvesting of the manioc root, meaning two or three years respectively, depending on the fertility of the soil. There are cases where only a single cycle of manioc planting is done, in the case of particular bi-annual species, and fallow is then done for two years. Similarly, the shifting systems are normally abandoned after the productivity of some bi-annual and middle-cycle crops is reduced, or after they reach senescence, three to five years after the planting. Additionally, there are also long cycle species, mainly orchard (fruit) species, which reach their production phenophase several years after the fallow portion of land is abandoned—corresponding to the beginning of the regeneration-succession phase of the secondary forest—, through the arrival of seed species coming from the surrounding forest or from previously existing ones in the soil seed bank, as long as undisturbed forest areas are predominant among the vegetation patch where the fallow portions of soil get to be established. This pattern of regeneration has also previously been reported in the Atlantic Forest (Gandolfi et al. 2007).

However, it is a general tendency that shifting systems in the tropical regions, established mainly by traditional settlements, have a short duration time of use for the purpose of taking advantage of the annual and bi-annual crops (between 2 and 4 years), and after the fallow areas having been abandoned, the resting time for the regeneration of the secondary forests is relatively long, longer than 25 years (Huber and Zent 1995; Villa et al. 2012). In contrast, Arroyo-Kalin (2012) presume that Pre-Columbian shifting cultivation might have been a lot more intense, based on short periods of fallow, with greater dependency on fire, and with more demanding labor requirements. Hence, it is important to highlight that an important trait of shifting systems has been their high agricultural biodiversity; even if as a general pattern in the basin, corn and sweet or bitter manioc are more frequently used in rotations (Jakovac et al. 2016). If the plantation of these crops constitutes the main reason for deforestation in the Amazon, then it would be very important to examine the regional history of the use of the land, in order to try to better understand the current patterns of the use of land and biodiversity (Arroyo-Kalin 2012; Roosevelt 2013).

Currently, agroforestry systems in protected areas in Brazil represent an important alternative that links the environmental, social and economic components, to program and maintain sustainable development in areas designated as "permanent preservation" and "legal reserve" through the creation of productive areas in small properties (Schroth et al. 2006; Oliveira-Neto 2012). In these Amazonian conservation units, agroforestry as a sustainable model of land use has the advantage of having high biodiversity (trees, shrubs, annual crops), and local resources to the same extent that other areas are protected in native forests or forests with different stages of succession regeneration (secondary forests). These systems are also important for the conservation of biodiversity and the conservation of carbon sinks as an environmental service with regional and global impacts. In addition, small farmers also have the option of harvesting agroforestry products to contribute to food security at the family and local levels, as well as the possibility of generating income through food marketing.

In accordance with Brazilian legal regulations, and on the basis of technical - environmental criteria of each type of protected area, in addition to knowing the different criteria for the adoption of agroforestry models, ARAFS are proposed for areas of permanent preservation (Oliveira-Neto et al. 2012). These systems are also known by other names such as "Forest Garden" or "Capoeiras melhoradas"; however, in the areas of the legal reserve, the adoption of the agrosilvicultural model known as "Taungya" (Nair 1993; Schroth et al. 2006; Vieira et al. 2009; Oliveira-Neto et al. 2012) is proposed. In general, the two proposed agroforestry models can have a considerable impact in relation to 1) conservation and recovery of carbon stocks above- and below-ground by increasing the density of tree planting, 2) production and sustainable management of non-timber forest products, 3) conservation of biodiversity through the succession of forest regeneration (passive restoration and rehabilitation of degraded areas). The main reasons for adoption of SAFs in protected areas are of an environmental nature that may have repercussions in the social and economic fields.

CONCLUSION

There is sufficient evidence explaining how different biophysical and anthropogenic drivers directly influence deforestation in the Amazon, according to the analysis of environmental and social processes such as climate change, population growth, immigration and colonization of new indigenous and non-indigenous settlements via new road networks, and thus, through a larger expansion of the basin's agricultural frontier, given the continuing demand for goods and services for domestic and commercial purposes.

Conservation and management of biodiversity and carbon stocks in Amazonian forests are important tools that can mitigate negative effects on the global climate. Therefore, the basic premise is that in order to implement REDD+ activities, including the adoption of agroforestry systems as a strategy for vegetation recovery in degraded landscapes it is necessary to identify edaphic and climatic indicators as principal environmental filters for the structure of vegetation communities, besides describing indicators of the dynamics of human-modified landscapes in Amazonia.

The proposed criteria permit an interaction among the diverse REDD+ activities, firstly through controlling emissions reduction through forest management, and secondly through an increase in vegetation biomass, which is an important carbon reserve, and thirdly, through the possibility of conservation and sustainable management of biodiversity.

The advance of REDD+ actions can be enhanced through the adoption of agroforestry systems as an alternative to shifting cultivation. These systems represent a strategy that can integrate different REDD+ activities. This intelligent agriculture model could contribute to a reduction in the necessity to deforest new areas and to the recuperation of degraded forests. It could also improve the practice of fallowing turning these areas productive, thus facilitating passive restoration during natural regeneration and processes of succession of secondary forest. These agroforestry systems would simultaneously contribute to conservation and recuperation of carbon stocks, and consequently to the maintenance of ecosystem services.

ACKNOWLEDGMENTS

The authors wish to acknowledge support from the Organization of American States (OAS) and the Brazilian Coordination for the Improvement of Higher Education Personnel (CAPES).

REFERENCES

Aguiar APD, Vieira ICG, Assis TO, Dalla-Nora E, Toledo PM, Santos-Junior RAO, Batistella M, Coelho AS, Savaget EK, Aragão LEOC, Nobre CA, Ometto JPH (2016) Land use change emission scenarios: anticipating a forest transition process in the Brazilian Amazon. *Glob Change Bio* 22:1821-1840.

Andela N, Morton DC, Giglio L (2017) A human-driven decline in global burned area. *Science* 356:1356-1362.

Anderson-Teixeira KJ, Wang MMH, Mcgarvey JC, Lebauer DS (2016) Carbon dynamics of mature and regrowth tropical forests derived from a pantropical database (TropForC-db). *Glob Change Biol* 22:1690-1709.

Angelsen A, McNeill D (2013) Evolución de REDD+. [Evolution of REDD+.] En Angelsen A, Brockhaus M, Sunderlin WD, Verchot LV (eds.) *Análisis de REDD+: Retos y opciones. CIFOR*, Bogor, Indonesia. pp 35-56.

Aragão LEOC, Poulter B, Barlow J (2014) Environmental change and the carbon balance of Amazonian forests. *Biol Rev* 89:913-931.

Arroyo-Kalin M (2012) Slash-burn-and-churn: Landscape history and crop cultivation in pre-Columbian Amazonia. *Quatern Int* 249:4-18.

Arroyo-Rodriguez,V, Roes M, Escobar F, Melo FPL, Santos BA, Tabarelli M, Chazdon R (2013) Plant beta-diversity in fragmented rain forests: testing floristic homogenization and differentiation hypotheses. *J Ecol* 101:1449-1458.

Arroyo-Rodríguez V, Melo FPL, Martínez-Ramos M, Bongers F, Chazdon RL, Meave JA, Norden N, Santos BA, Leal IR, Tabarelli M (2015)

Multiple successional pathways in human-modified tropical landscapes: new insights from forest succession, forest fragmentation and landscape ecology research. *Biol. Rev* 92:326-340.

Bachelet KD, Forrest M, Lasslop G et al. (2017) A human-driven decline in global burned area. *Science* 356:1356-1362.

Baiser B, Olden JD, Record S, Lockwood JL, McKinney ML (2012) Pattern and process of biotic homogenization in the New Pangaea. *Proc R Soc B Biol Sci* 279:4772-4777.

Barlow J, Lennox GD, Ferreira J et al. (2016) Anthropogenic disturbance in tropical forests can double biodiversity loss from deforestation. *Nature* 535:144-147.

Barton H, Denham T, Neumann K, Arroyo-Kalin M (2012) Long-term perspectives on human occupation of tropical rainforests: An introductory overview. *Quatern Int* 249:1-3.

Betts RA, Malhi Y, Roberts JT (2008) The future of the Amazon: new perspectives from climate, ecosystem and social sciences. *Philos T R Soc* B 363:1729-1735.

Booth BBB, Jones CD, Collins M et al. (2012) High sensitivity of future global warming to land carbon cycle processes. *Environ Res Lett* 7:024002.

Bush MB, McMichael CH, Piperno DR, Silman MR, Barlow J, Peres CA, Power M, Palace MW (2015) Anthropogenic influence on Amazonian forests in pre-history: An ecological perspective. *J Biogeogr* 42:2277-2288.

Brando PM, Balch JK, Nepstad DC, Morton DC, Francis EP, Michael TC, Silvério D, Macedo MN, Davidson EA, Nóbrega CC, Alencar A, Soares-Filho BS (2014) Abrupt increases in Amazonian tree mortality due to drought–fire interactions. *Proc Natl Acad Sci* 111:6347-6352.

Brienen RJW, Phillips OL, Feldpausch TR et al. (2015) Long-term decline of the Amazon carbon sink. *Nature* 519:344-360.

Boucher D, Roquemore S, Fitzhughl E (2013) Brazil's Success in Reducing Deforestation. Trop Conserv Sci. 6:426-445.

Cardoso D, Sarkinen T, Alexander S, Amorim, AM, Bittrich V, Celis M, Forzza, RC (2017) Amazon plant diversity revealed by a taxonomically verified species list. *Proc Natl Acad Sci* 114:10695-10700

Cerri CEP, Easter M, Paustian K, Killian K, Coleman K, Bernoux M, Falloon P, Powlson DS, Batjes NH, Milne E, Cerri CC (2007) Predicted soil organic carbon stocks and changes in the Brazilian Amazon between 2000 and 2030. *Agr Ecosyst Environ* 122:58-72.

Chazdon RL (2014) *Second growth: The promise of tropical forest regeneration in an age of deforestation.* University of Chicago Press, Chicago, Illinois, USA.

Devaraju N, Bala G, Modak A (2015) Effects of large-scale deforestation on precipitation in the monsoon regions: Remote versus local effects. *Proc Natl Acad Sci* 112:3257-3262.

Derroire G, Balvanera P, Castellanos-Castro C et al. (2016) Resilience of tropical dry forests – a meta-analysis of changes in species diversity and composition during secondary succession. *Oikos* 125:1386-1397

Diniz FH, Hoogstra-Klein MA, Kok K, Arts B (2013) Livelihood strategies in settlement projects in the Brazilian Amazon: Determining drivers and factors within the Agrarian Reform Program. *J Rural Stud* 32:196-207.

Dent DH, DeWalt SJ, Denslow JS (2013) Secondary forests of central Panama increase in similarity to old-growth forest over time in shade tolerance but not species composition. *J Veg Sci* 24:530-542.

D'Oliveira MV, Alvarado EC, Santos JC, Carvalho JA (2011) Forest natural regeneration and biomass production after slash and burn in a seasonally dry forest in the Southern Brazilian Amazon. *For Ecol Manag* 261:1490-1498.

Engel VL, Parrotta JA (2001) An evaluation of direct seeding for reforestation of degraded lands in central Sao Paulo state, Brazil. *For Ecol Manage* 152:169-181.

Espindola GM, Aguiar APD, Pebesma E, Câmara G, Fonseca L (2012) Agricultural land use dynamics in the Brazilian Amazon based on remote sensing and census data. *Appl Geogr* 32:240-252

FAO (2010) Global Forest Resources Assessment 2010. FAO Forestry Paper 163. Food and Agriculture Organization of the United Nations. FAO (2013) *Intelligent Agriculture Climatic*. 84 p.

Fauset S, Johnson MO, Gloor M et al. (2015) Hyperdominance in Amazonian forest carbon cycling. *Nat Commun* 6:6857. DOI:10.1038/ncomms7857.

Feldpausch TR, Rondon MA, Fernandes ECM, Riha SJ, Wandelli E (2004) Carbon and Nutrient Accumulation in Secondary Forests Regenerating on Pastures in Central Amazonia. *Ecol Appl* 14:S164-S176.

Feldpausch TR, Lloyd J, Lewis SL et al. (2012) Tree height integrated into pantropical forest biomass estimates. *Biogeosciences* 9:3381-3403.

Finegan B. (1996) Pattern and process in neotropical secondary rain forests: The first 100 years of succession. *Trends Ecol Evol* 11:119-124.

Gandolfi S, Rodrigues RR, Martins SV (2007) Theoretical bases of the forest ecological restoration. In Rodrigues RR, Martins SV, Gandolfi S (eds.). *High Diversity Forest Restoration in Degraded Areas: Methods and Projects in Brazil*. Nova Science Publishers, Inc. New York, USA. pp 27-59.

Gibson L, Lee TM, Koh LP et al. (2011) Primary forests are irreplaceable for sustaining tropical biodiversity. *Nature* 478:378-383.

Goulart AC, Macario K, Scheel-Ybert R et al. (2017) Charcoal chronology of the Amazon forest: A record of biodiversity preserved by ancient fires. *Quatern Geochr* 41:180-186.

Guariguata MR, Ostertag R (2001) Neotropical secondary forest succession: changes in structural and functional characteristics. *For Ecol Manage* 148:185-206.

Houghton RA, Skole DL, Nobre CA, Hackler JL, Lawrence KT, Chomentowski WH (2000) Annual fluxes of carbon from deforestation and regrowth in the Brazilian Amazon. *Nature* 403:301-304.

Huber O, Zent S (1995) Indigenous people and vegetation in the Venezuela Guayana: Some ecological considerations. *Scientia Guaianae* 5:37-64.

Huntingford C, Zelazowski P, Galbraith D et al. (2013) Simulated resilience of tropical rainforests to CO_2-induced climate change. *Nature Geosci* 6:268-273.

IPCC (Intergovernmental Panel on climate change), 2000. IPCC. Special Report on Land Use, Land-Use Change and Forestry. *A special report of the intergovernmental Panel on climate change.* Approved at IPCC Plenary (Montreal May 2000). IPCC, world Meteorological Organization, Geneva, Switzerland.

Jagger P, Lawlor K, Brockhaus M, Gebara MF, Sonwa DJ, Resosudarmo IAP (2013) Salvaguardas de REDD+ en el discurso de políticas nacionales y proyectos piloto. [REDD + safeguards in the discourse of national policies and pilot projects.] En Angelsen A, Brockhaus M, Sunderlin WD, Verchot LV (eds.) *Análisis de REDD+: Retos y opciones.* CIFOR, Bogor, Indonesia. pp. 340-358.

Jakovac CC, Peña-Claros M, Kuyper TW, Bongers F (2015) Loss of secondary-forest resilience by land-use intensification in the Amazon. *J. Ecol* 103:67-77.

Jakovac CC, Peña-claros M, Mesquita R, Bongers F, Kuyper TW (2016) Swiddens under transition: consequences of agricultural intensification in the Amazon. *Agr Ecosyst Environ* 218:116-125.

Kapos V, Kurz WA, Gardner T (2012) Impacts of forest and land management on biodiversity and carbon. In Parrota JA, Wildburger C, Mansourian S (eds.) *Understanding Relationships between Biodiversity, Carbon, Forests and People: The Key to Achieving REDD+ Objectives. Vienna, Austria. A global assessment report IUFRO.* pp. 53-78.

Kato OR (2006) Uso de agroflorestais no manejo de florestas secundarias. [Use of agroforestry in the management of secondary forests.] In: Gama-Rodrigues AC, Barros NF, Gama-Rodrigues EF (eds.) *Sistemas agroflorestais: bases cientificas para o desenvolvimento sustentável.* Universidade do Norte Fluminense, Brasil. Pp. 119-139.

Klanderud K, Mbolatiana HZH, Vololomboahangy MN et al. (2010) Recovery of plant species richness and composition after slash-and-burn agriculture in a tropical rainforest in Madagascar. *Biodivers Conserv* 19:187-204.

Laurance WF, Cochrane MA, Bergen S, Fearnside PM, Delamonica P, Barber C, DAngelo S, Fernandes T (2001) The future of the Brazilian Amazon. *Science* 291:438-439.

Laurance WF (2002) Predictors of deforestation in the Brazilian Amazon. *J Biogeogr* 29:737-748.

Laurance WF, Albernaz AKM, Fearnside FM, Vasconcelos HL, Ferreira LV (2004) Deforestation in Amazonian. *Science* 21:1109-1111.

Lawrence D, Vandecar K (2015) Effects of tropical deforestation on climate and agriculture. *Nat Clim Change* 5:27–36.

Luyssaert S, Schulze ED, Borner A, Knohl A, Hessenmoller D, Law BE, Ciais P, Grace J (2008) Old-growth forests as global carbon sinks. *Nature* 455:213-215.

Levis C, Costa FRC, Bongers F et al. (2017) Persistent effects of pre-Columbian plant domestication on Amazonian forest composition. *Science* 355:925-931.

Lewis SL, Edwards DP, Galbraith D (2015) Increasing human dominance of tropical forests. *Science* 349:827-832.

Llambí L, Llambí LD (2000) A Transdiciplinary Framework for the Analysis of Tropical Agroecosystems Transformations: The advance of the agricultural frontier in Venezuela`s Orinoco/Amazon Region." In: Higgins V, Lawrence G, Lockie S (eds.). *Environment, Society and Natural Resource Management: theoretical perspectives from Australasia and the Americas.* Edward Elgar Publishing, Cheltenham, UK. pp. 53-70.

Malhi Y, Wood D, Baker TR (2006) The regional variation of aboveground live biomass in old-growth Amazonian forests. *Glob Change Biol* 12:1107-1138.

Martin PA, Newton AC, Bullock JM (2013) Carbon pools recover more quickly than plant biodiversity in tropical secondary forests. *Proc R Soc B* 280:20132236.

Martorano LG, Siviero MA, Tourne DCM, Vieira SB, Fitzjarrald DR, Vettorazzi CA, Brienza JS, Yeared JAG, Meyering E, Lisboa LSS (2016). Agriculture and forest: A sustainable strategy in the Brazilian Amazon. *Australian J crop Science* 10:1136-1143.

Martorano LG, Lisboa LS, Villa PM, Moraes JR (2017) Fragilidade das Terras pelo processo erosivo das chuvas em áreas antrópicas e declivosas na Amazônia Legal. [Fragility of the Lands by the erosive process of the rains in anthropic and sloping areas in the Legal Amazon.] In: *XXXVI Congresso Brasileiro de Ciência do Solo*. Belém, Pará.

Mitchard ETA, Feldpausch TR, Brienen RJW et al. (2014) Markedly divergent estimates of Amazon forest carbon density from ground plots and satellites. *Glob Ecol Biogeogr* 23:935-946.

Meiners SJ, Cadotte MW, Fridley JD, Pickett STA, Walker LR (2015) Is successional research nearing its climax? New approaches for understanding dynamic communities. *Funct Ecol* 29:154-164.

McMichael CNH, Matthews-Bird F, Farfan-Rios W, Feeley KJ (2017) Ancient human disturbances may be skewing our understanding of Amazonian forests. *Proc Natl Acad Sci* 114:1-6.

Nair PKR (1993) *An Introduction to Agroforestry*. Kluwer Academic Publishers. Dordrecht. 489 p.

Nair PKR, Kumar BM, Nair VD (2009) Agroforestry as a strategy for carbon sequestration. *J Plant Nutr Soil Sci* 172:10-23.

Nair PKR, Nair VD, Kumar BM, Showalter JM (2010) Carbon sequestration in agroforestry systems. *Adv Agron* 108:237-307.

Nepstad D, McGrath D, Stickler C et al. (2014) Slowing Amazon deforestation through public policy and interventions in beef and soy supply chains. *Science* 344:1118-1123.

Nobre CA, Sampaio G, Bormac LS, Castilla-Rubio JC, Silvae JS, Cardoso M (2016) Land-use and climate change risks in the Amazon and the need of a novel sustainable development paradigm. *Proc Natl Acad Sci* 113:10759-10768.

Nogueira EM, Yanai AM, Fonseca FOR, Fearnside PM (2015) Carbon stock loss from deforestation through 2013 in Brazilian Amazonia. *Glob Change Biol* 21:1271-1292.

Norden N, Chazdon RL, Chao A, Jiang Y, Vílchez-Alvarado B (2009) Resilience of tropical rain forests: tree community reassembly in secondary forests. *Ecol Lett* 12:385-394.

Olden JD, Rooney TP (2006) On defining and quantifying biotic homogenization. *Glob Ecol Biogeogr* 15:113-120.

Oliveira Neto SN (2012) Sistemas Agroflorestais para adequação ambiental de propriedades rurais. [Agroforestry systems for environmental adaptation of rural properties.] *Informe Agropecuário* 33:70-77.

Oliver TH, Heard MS, Isaac NJB et al. (2015) Biodiversity and Resilience of Ecosystem Functions. *Trends Ecol Evol* 30:673-684.

Palomeque X, Günter S, Siddons D et al. (2017) Natural or assisted succession as approach of forest recovery on abandoned lands with different land use history in the Andes of Southern Ecuador. *New Forests* 48:643-662.

Pascarella JB, Aide TM, Serrano MI, Zimmerman JK (2000) Land-use history and forest regeneration in the Cayey Mountains, Puerto Rico. *Ecosystems* 3:217-228.

Piperno DR (2009) Identifying crop plants with phytoliths (and starch grains) in Central and South America: a review and an update of the evidence. *Quatern Int* 193:146-159.

Piperno DR, McMichael C, Bush MB (2015) Amazonia and the Anthropocene: What was the spatial extent and intensity of human landscape modification in the Amazon Basin at the end of prehistory? *Holocene* 25:1-10.

Poorter L, Bongers F, Aide TM et al. (2016) Biomass resilience of Neotropical secondary forests. *Nature* 530:211-217.

Poorter L, van der sande M, AJMM Arets et al. 2017. Biodiversity and climate determine the functioning of Neotropical forests. *Glob Ecol Biogeogr* 26:1423-1434.

Porro R, Miller RP, Tito MR, Donovan JA, Vivan JL, Trancoso R, Van Kanten RF, Grijalva JE, Ramirez BL, Gonçalves AL (2012) Agroforestry in the Amazon Region: A Pathway for Balancing Conservation and Development. In Nair PKR, Garrity D (eds.). Agroforestry: The Future of Global Land Use. *Advances in Agroforestry* 9:391-428.

Roosevelt AC (2013) The Amazon and the Anthropocene: 13,000 years of human influence in a tropical rainforest. *Anthropocene* 4:69-87.

Rodrigues-Filho S, Verburg R, Bursztyn M, Lindoso D, Debortoli N, Vilhena AMG (2015) Election-driven weakening of deforestation control in the Brazilian Amazon. *Land Use Policy* 43:111-118.

Saatchi SS, Harris NL, Brown S et al. (2011) Benchmark map of forest carbon stocks in tropical regions across three continents. *Proc Natl Acad Sci* 108:9899-9904.

Silvério DV, Brando PM, Macedo MN, Beck PSA, Bustamante M, Coe MT (2015) Agricultural expansion dominates climate changes in southeastern Amazonia: the overlooked non-GHG forcing. *Environ Res Let.* 10:104015.

Soares-Filho B, Moutinho P, Nepstad D (2010) Role of Brazilian Amazon protected areas in climate change mitigation. *Proc Natl Acad Sci* 107:10821-10826.

Schroth G, Souza M, Jerozolimski A (2006) Agroforestry and the conservation of forest and biodiversity in tropical landcapes- on-site and off-site effects and synergies with environmental legistation. In Gama-Rodrigues AC, Barros NF, Gama-Rodrigues EF (eds.) *Sistemas agroflorestais: bases científicas para o desenvolvimento sustentável. Fluminense,* Brasil. Universidade do Norte. pp. 67-86.

Solar RR; Barlow J, Ferreira J et al. (2015) How pervasive is biotic homogenization in human-modified tropical forest landscapes? *Ecol Lett* 18:1108-1118.

Solar RR, Barlow J, Andersene AN, Schoereder JH, Berenguer E, Ferreira JN, Gardner TA (2016) Biodiversity consequences of land-use change and forest disturbance in the Amazon: A multi-scale assessment using ant communities. *Biol Conserv* 197:98-107.

Sougata B, Shibu J (2012) Agroforestry for biomass production and carbon sequestration: an overview. *Agrof Syst* 86:105-111.

Strassburg BBN, Kelly A, Balmford A, Davies RG, Gibbs HK, Lovett A, Miles L, Orme CDL, Price J, Turner RK, Rodrigues ASL (2010) Global congruence of carbon storage and biodiversity in terrestrial ecosystems. *Conserv Lett* 3:98-105.

ter Steege H, Pitman NCA, Sabatier C et al. (2013) Hyperdominance in the Amazonian tree flora. *Science* 342:1243092.

Thompson ID, Ferreira J, Gardner T, Guariguata M, Koh LP, Okabe K, Pan Y, Schmitt CB, Tylianakis J (2012) Forest biodiversity, carbon and other ecosystem services: relationships and impacts of deforestation and forest degradation. In Parrota J, Wildburger C, Mansourian S (eds.) *Understanding Relationships between Biodiversity, Carbon, Forests and People: The Key to Achieving REDD+ Objectives.* Vienna, Austria. A global assessment report IUFRO. Pp. 21-52.

Vieira DLM, Holl KD, Peneireiro FM (2009) Agrosucessional restoration as s strategy to facilitate tropical forest recovery. *Rest Ecol* 17:451-459.

Villa PM, Riera A, Mota N, Belandria A, Camacho D, Sánchez I, Infante J, Oliveros G, Delgado L, García J (2012) *Agricultura Piaroa en la cuenca del río Cataniapo, estado Amazonas: Un enfoque agroecológico.* [*Piaroa agriculture in the Cataniapo river basin, Amazonas state: An agro-ecological approach.*] Fundación PROBIODIVERSA, PPD/GEF /PNUD, INIA. 46 p.

Villa PM, Martins SV, Monsanto L, de Oliveira Neto SN, Cancio NM (2015) La agroforestería como estrategia para la recuperación y conservación de reservas de carbono en bosques de la Amazonía. [Agroforestry as a strategy for the recovery and conservation of carbon stocks in forests of the Amazon.] *Bosque* 36:347-356.

Villa PM, Martins SV, Oliveira Neto SN, Rodrigues AC (2017) Anthropogenic and biophysical predictors of deforestation in the Amazon: towards integrating REDD+ activities. *Bosque* 38:433-446.

Villa PM, Martins SV, Oliveira Neto SN, Rodrigues AC, Vieira N, Delgado L, Cancio NM (2018a) Woody species diversity as an indicator of the forest recovery after shifting cultivation disturbance in the northern Amazon. *Ecol Indic* 95: 687-694

Villa PM, Martins SV, Oliveira Neto SN, Rodrigues AC, Martorano L, Delgado L, Cancio NM, Gastauerg M (2018b) Intensification of shifting cultivation reduces forest resilience in the northern Amazon *For Ecol Manage* 430: 312-320.

Yanai AM, Nogueira EM, Alencastro Graça PML, Philip Fearnside M (2017) Deforestation and Carbon Stock Loss in Brazil's Amazonian Settlements. *Environ Manage* 59:393-409.

In: Forest Conservation
Editor: Pedro V. Eisenlohr

ISBN: 978-1-53614-559-5
© 2019 Nova Science Publishers, Inc.

Chapter 7

FOREST CONSERVATION AND ITS CHALLENGES IN TROPICAL AFRICA

Tsegaye Tagesse Gatiso[*]
Max Planck Institute for Evolutionary Anthropology,
Leipzig, Germany

INTRODUCTION

Forest resources play a crucial role in sustaining human life. Human beings depend on forest resources for food, clean water, air and other life sustaining basic needs. For example, forest resources contribute with a significant proportion of the livelihoods of rural population in the developing world (Adhikari 2005; Sunderlin et al. 2005; Mamo et al. 2007; Babulo et al. 2008; Appiah et al. 2009; Gatiso and Wossen 2014; Gatiso 2017). In tropical Africa typical forest dependent rural households earn 28-38% of their annual household income from forest resources (Mamo et al. 2007; Babulo et al. 2008; Appiah et al. 2009; Gatiso and Wossen 2014;

---

[*] Corresponding Author Email: tsegaye_gatiso@eva.mpg.de.

Gatiso 2017). In addition, forest resources also serve as assets for poor households to earn their livelihoods (Narian et al. 2005).

Nevertheless, the degradation and fragmentation of forests have continued around the globe, particularly in the tropical areas, unabated (FAO 2015). Approximately 9.4 million hectares of tropical forest are lost annually (FAO 2000).

Almost half of the world's forest resource is lost in the past 800 years (Sunderlin et al. 2004). Forest degradation is contributing almost 25% of CO_2 emissions (FAO 2005). Recent estimates show that tropical deforestation and forest degradation contribute about 10% of the global greenhouse gas emissions (IPCC 2013).

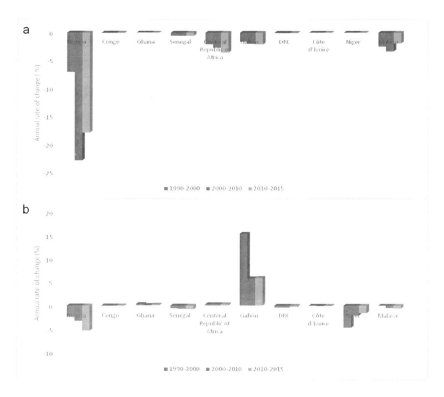

Figure 1. Annual rate of change in primary and other naturally regenerating forest cover from 1990 to 2015 in selected African countries (FAO 2015). a. Primary forest change; b. Other naturally regenerated forest change rate.

As shown in Figure 1 the overwhelming majority of the Sub-Saharan African countries have shown a decline in both primary and other naturally regenerating forest cover from 1990-2015.

In response to the alarming rate of deforestation and forest degradation, sustainable management of forest resources has remained the top agenda for conservationists, development agents, policy makers, researchers, governments, NGOs, among others.

In this chapter, emphasis will be given to the historical background of forest conservation policies in tropical areas, the relationship between forest degradation and poverty, and challenges to forest conservation in tropical areas.

Forest Conservation Policies in the Tropics: Past and Present

In the race against time to curtail the drastic loss of forest resources in tropical areas, different management approaches have been experimented with. These approaches can broadly be categorized as command and control, and participatory approaches. In the following section we discuss these approaches in detail.

The Centrally Controlled Protectionist (Or Command and Control) Approach

This approach is one of the oldest and widely used approaches in forest conservation in developing countries. It dominantly follows the philosophy of "fence and fines," and envisages to conserve forest through centralized-regulatory system. It aims at preventing illegal forest use through monitoring and punishments. The proponents of this approach widely argue for top-down forest management, and claim that sustainable forest conservation is only possible when it is managed by governments (Terborgh 1999). They argue that the bottom-up approaches may channel away a significant portion of the already limited conservation funding to other activities that are not directly linked to conservation. In the face of the current alarming rate of population growth around forests, coupled with more access to modern technology and modern life style, the bottom-up approaches (e.g., Community-Based Conservation) may not successfully protect forest resources in tropical areas (Kramer and van Schaik 1997).

Nonetheless, the strict protectionist policies have been reported to be ineffective in conserving forest resources in the tropics (Salafsky and Wollenberg 2000; Brown 2002; Andrade and Rhodes 2012). The ineffectiveness of this approach for forest conservation in tropical areas could be due to four major reasons. First, it is very expensive to effectively enforce law to deter illegal use. The problems are also compounded by lack of technical expertise and financial resources by the departments responsible for forest conservation. Moreover, the success of the "fence and fines" approaches depends on the effectiveness of the guarding system. But the least-paid and limited number of guards in most African countries lack the incentive to defend the forest from illegal encroachments. Second, centrally controlled policies may lead to a shift in responsibility from individuals to the regulatory agencies, and make forest adjacent communities more self-centered and less social-oriented (e.g., Ostman 1998; Ostrom 1999; Cardenas 2000). Enforcement of externally imposed rules may also discourage the formation of social norms among the local communities (Ostrom 1999). Third, externally imposed rules are less likely to be accepted by local communities and hence, villagers tend to have negative attitude towards forest conservation in general and management authorities in particular (Nepal and Weber 1995). Four, strict protectionist policies may trigger adverse social impacts on local communities by disrupting their traditional ways of living and limiting their control of and access to natural resources (Wilshusen et al. 2001). Fifth, local communities may associate externally imposed rules with bad historical experience of colonial domination or state coercion.

PARTICIPATORY APPROACHES

Since the early 1970s the focus of conservation policies has shifted towards approaches that involve local communities in natural resource conservation (Charnley and Poe 2007). Four major reasons could be cited as reasons for this shift in the forest conservation policy regime. First, as discussed above the strict protectionist approach was reported to be ineffective to conserve forest resources in most developing countries. Second, the inclusionary approaches pave the way to reconcile the two sides of the conservation equation: people and nature. Further, they create an opportunity to achieve the two major goals of sustainable development: conservation and development. Third, as local communities are in a better position to conserve natural resources around them (Thakadu 2005), the involvement of local communities in rule making, monitoring and enforcement may reduce the cost of conservation (Rahut et al. 2015). It makes it easier for conservation authorities to target illegal and unsustainable resource uses. At the same time, it makes it difficult for illegal loggers to use local villagers as accomplices to elude arrest and incarcerations. Fourth, community involvement may increase the rate of compliance of locals with the rules and regulations of conservation areas (Andrade and Rhodes 2012). In their meta-analysis, Andrade and Rhodes (2012) assert that community participation in the decision making process of the protected areas increases the rate of compliance with the regulations of the protected areas. Persha et al. (2011) also report that the level of local communities' participation in decision-making process and conservation outcomes are strongly and positively related. Thus, according to the proponents of participatory approaches, allowing the local communities to participate in nature conservation enables forest conservation authorities usually intertwined with shortage of resources to effectively conserve forest resources at low cost.

The participatory forest management policies have been implemented in different parts of the world with different names, which could broadly be categorized into Integrated Conservation and Development Programs (ICDP), and Community-Based Conservation Programs (CBCP), Joint forest management programs (JFMP), Payment for Ecosystem Services (PES). Both CBCP and ICDPs aim at providing economic incentives to local communities to conserve forests around them. However, they have a major difference in the approach they use. While the former emphasizes sustainable use of natural resources and grater decision making power devolvement to the local community (Bandyopadhyay and Tembo 2010; Spiteri and Nepal 2006), the later focuses on generating alternatives to resource use and compensates for reduction in resource use (Brandon and Wells 1992). The idea of community-based natural resource management follows the seminal work of Elinor Ostrom on the capacity of local resource users to self-govern natural resources under certain circumstances such as clearly defined boundaries, control over the resource, dependence on the resource, presence of appropriate rules and regulation and others (Ostrom 1990). Joint-forest management programs involve the partnership and collaboration between the state and the local communities to manage forest resources, and hence it is also known as co-management of forests by the state and local communities. On the other hand, PES is a market based approach that involves suppliers and buyers of ecosystem/ environmental services, and the suppliers are paid for the environmental services they provide by the buyers of the services. Usually, local communities are the suppliers and governments, NGOs, and conservation agencies are the buyers of the environmental services. PES is relatively the newest approach, and it seeks to entice local communities to change their resource use behavior in an environment friendly way using economic incentives, and local communities are paid for producing the desired environmental service. Since 2007 PES has become part of the famous UN program called REED+ (reducing emissions from deforestation and forest degradation). REDD+ is a type of PES that involves a set of actions aimed at reducing emissions from deforestation and forest degradation in developing countries (Angelsen 2009). REDD+ is a multi-stage performance-based PES, where the beneficiaries are paid based on their contribution to reduction in deforestation and forest degradation (i.e., based on their performance). In the first stage, specific countries would be paid for documented emission reductions. In the second stage, forest owners and

Forest Conservation and Its Challenges in Tropical Africa 213

users within these countries would be paid for their contribution to the achievement of the specified national goal. The payments for the forest owners at the second stage could be made at individual or group level; which shows that REED+ could sometimes combine both CBCP and PES (Angelsen 2009).

It should be noted that the participatory approaches (be it CBCP or ICDP or PES) are not without limitations either. For example, Spiteri and Nepal (2006) reported that CBCP in some parts of Nepal were not as effective as expected. ICDP are also reported to be less effective in stemming natural resource degradation in some contexts (Wells et al. 1999; Barrett and Arcese 1995) and are mainly criticized for their failure to make direct connection between conservation and the development programs offered (Ferraro and Kiss 2002). Further, in many tropical areas, the implementation of PES may be impaired by lack of efficient bureaucracy, clear property rights, market for ecosystem services, and high transaction costs among others (Wunder 2013; Gatiso et al. 2018).

FOREST RESOURCES AND RURAL POVERTY IN THE TROPICS

Tropical forests host the majority of biodiversity hotspots of the world. Unfortunately, these areas also represent the poorest segment of the world's population (MEA 2005), particularly the rural areas where people substantially depend on forests for their livelihood. In relative terms, the poorest segment of the rural population heavily depends on forest resources than the better-off segment (Lopez-Feldman et al. 2011).

Forest resources play a crucial role in supporting the livelihood of the rural households and reducing poverty (e.g., Reddy and Chakravarty 1999; Cavendish 1999; Cavendish 2003; Fisher 2004; Vedeld et al. 2004, Gatiso and Wosen 2014; Gatiso 2017). Limiting access to forest resources in the name of conservation could lead to a substantial reduction of the livelihood of rural poor and push them into further poverty. For instance, in the absence of forest income, the incidence of rural poverty may increase by about 28% (Reddy and Chakravarty 1999).

Furthermore, it has widely been documented that forest products play a crucial role in reducing inequality in income distribution among rural communities. For instance, Cavendish (1999) found that in rural Zimbabwe the situation of inequality may be overstated by as much as 44%, depending on the measure used in the calculation of inequality. Fisher (2004) reported that if forest income is excluded from the inequality calculation, the income inequality increases by 12% in the southern region of Malawi. Gatiso and Wossen (2014) also found that rural inequality may increase by 24% had the households from Southwest Ethiopia had no access to forest resources.

On the other hand, studies show that households in natural resource rich areas, particularly in developing countries, are often poor (MEA 2005). Furthermore, majority of the rural poor live in and around forest areas (World Bank 2003).

The relationship between poverty and forest dependence is a contentious issue. Some studies have shown that the level of dependence on forest resources declines with the level of income, implying a linear but negative relationship between them (Reddy and Chakravary 1999). On the other hand, others have found that the relationship between forest dependence and poverty could be non-linear, and forest dependence first declines with the level of income, reaches minimum and after a certain point it continuously rises with the level of income (Narrian et al. 2005; Lopez-Feldman et al. 2011). Though the role of forest resources in supporting the livelihoods of rural poor is unquestionable, there could also be a possibility for forest dependence to perpetuate poverty. Poverty leads people to depend more on forest resources, and more forest dependence leads to forest degradation, which in turn leads to more poverty as forest degradation leads to deterioration of the resource base of poor rural households. Consequently, to support their livelihood, the poor households harvest more resources, which further degrades the forest resources around them and vicious circle between forest dependency, forest degradation and poverty persists (Barret et al. 2011). This is commonly known as "poverty-environment trap."

CHALLENGES TO SUSTAINABLE FOREST CONSERVATION IN TROPICAL AFRICA

Despite the considerable effort and resources devoted to tropical forest conservation, the forest resource has continued to decline in tropical Africa. Why?

Burgeoning Population Growth and Poverty

The human population in tropical areas continues to grow at an alarming rate. In most countries, the growth rate is above 3%, while the world average is 1.3% (World Bank 2015a). At the same time, the forest resources in those countries continued to decline (see Figure 1). As the human population grows in the face of limited and declining forest resources, the forest related labor productivity declines. This is because the returns to the rural labor employed in forest related activities depend on the availability of the forest resource and the amount of rural labor engaged in forest related employment. Thus, the increase in one accompanied by a decline in the other, leads to a decrease in forest related incomes and, with it, standards of living falls and poverty increases. In response to increased poverty, people extract more forest resources, which in turn leads to forest degradation further reinforcing rural poverty, and the environment-poverty trap sets-off.

Further, more population means more mouths to feed and more demand for food, and more need for agricultural land, given lack of mechanization in most tropical areas. Often, the expansion of agricultural land is done by clearing more forests. High population growth also leads to high population density over a given plot of land, causing more farm fragmentation and low productivity. Human density estimates are significantly higher in tropical areas compared to the average world population density of 42 people/km2 (World Bank 2015b).

Exclusion of Local Communities from Decision Making Process of Forest Management

In the past, most conservation initiatives do not encourage the local communities to take part in conservation activities. As a result the local communities associate the conservation initiatives with the era of colonization and forceful control of resources by outsiders (Wilshusen et al. 2001). Most of the time conservation programs, even those that claim to be participatory, are designed by outsiders and lack contextualization to the local reality (Nuesiri 2015). Often, they fail to incorporate the wishes and aspirations of local communities in the decision making process (Treves et al. 2006; Andrade and Rhodes 2012). In most participatory approaches the local communities' participation is often passive (Colchester 2000; Brown 2002). Consequently, conservation intervention may lack popular support and social acceptance, which in turn makes them less effective in achieving their desired goal of conservation (Anthony 2007; Andrade and Rhodes 2012; Bennett and Dearden 2014).

Research, however, shows that local communities' participation in the decision making process of rural projects leads to higher satisfaction, and hence, higher willingness to contribute for the projects (Pouta et al. 2002; Olken 2010; Cavalcanti et al. 2010). Local communities are more likely to comply with the regulations of conservation areas when they are established through participation (Pretty and Smith 2004; Andrade and Rhodes 2012).

Limited Resources for Law Enforcement

Forest conservation departments in most developing countries are underfinanced and are constrained by limited resources. Moreover, in most of the cases the law enforcement staffs are underpaid and lack the incentive to put their maximum effort to strictly enforce the law. The low salary of rangers may also open a door for complicity in illegal activities and makes

Forest Conservation and Its Challenges in Tropical Africa   217

the rangers prone to bribe and corruption. For instance, at some point in Kenya one-third of the rhino losses was attributed to park staff (Bonner, 1993).

Public Goods Dilemma

The public goods dilemma is a situation that arises when the benefits of conservation accrue to the global community while the costs of conservation are disproportionately born by local communities. For centuries, the local communities have been living in and around forest areas, and extracting forest resources has been their way of life. Thus, taking the resource away and restricting or denying access to the resource in the name of conservation adversely affect the welfare of the local communities and aggravate rural poverty (e.g., West and Brechin 1991; Kapoor 2001; Narain et al. 2008, Babulo et al. 2008; Gatiso and Wossen 2014). Most of the time the benefits from forest conservation accrue to the global community, but the cost of conservation in terms of altering the traditional way of living of the indigenous and local communities is exclusively born by forest adjacent communities. Thus, local communities may become reluctant to sacrifice their livelihoods to promote the common interest of global communities (Wilshusen et al. 2001), which may lead to conflict of interest and resistance to conservation.

Corruption

One of the most important factors that predict the success of forest conservation in developing countries is the level of corruption in the country (Smith et al. 2003). Countries with higher corruption index are reported to be poorly performed in conserving natural resources (Smith et al. 2003). The success of local programs to achieve their goals could be due significantly hampered by elite capture (e.g., Bardhan 2002) and limited benefits passing onto the local communities (Pinaar and Krapeletswe 2005).

Armed Conflicts and Civil Unrest

Most of the countries in tropical areas are unstable and armed conflicts between different factions are so rampant. Armed conflicts contribute to deforestation by making forest management difficult. Governments may also order clearing of forests to deal with armed rebels as it was done in Uganda in the early 1980s (Winterbottom and Eilu 2006). In addition, armed conflicts and civil unrests may lead to internal displacement of people and influx of refugees from neighboring countries, which may lead to deforestation and degradation of natural resources as camps for displaced people or refugee are often established in previously unsettled areas. Moreover, constructing camps needs cutting trees. Refuges also add pressure on the forest resources for their household consumption such as firewood, farm implements, household utensils and others. Armed groups may also use logging as a source of finance to buy arms.

CONCLUSION

Forest resources play a crucial role in the livelihood of rural households in developing countries. Rural households directly depend on forest resources to earn their living, and substantial part of their income comes from forest related activities. Unfortunately, the forest resource has been on a down ward spiral and most tropical areas have been experiencing huge loss of forests. These areas are also homes to the largest proportion of the world's poor.

The relationship between poverty and forest degradation is complex. Often poor people depend more on forest resources, and their heavy dependence causes forest degradation, which in turn leads to more poverty as with forest degradation rural households lose their means of living. To support their livelihood they harvest more resources, which further degrades the forest resource, and this vicious circle is commonly known as the "poverty-environment trap."

In the past the command and control was the dominant approach used by policy makers in their attempt to avert forest degradation in tropical areas, though it was widely reported to be ineffective. Recently, most countries have started recognizing the crucial role of the forest adjacent communities in conserving forest, and have started implementing participatory approaches. These approaches vary from country to country, and they also differ based on their design. Mostly, the participatory approaches include CBCP, JCP, ICDP and PES. Regardless of the massive efforts to stem forest degradation in tropical areas, the resource continues to decline. The major challenges to tropical forest conservation could be high population growth, poverty, limited resources for law enforcement, corruption, armed conflicts and civil unrest etc. Thus, forest conservation policies in tropical Africa, to be effective in forest conservation, should account for these important challenges and have a plan to minimize their effects.

REFERENCES

Adhikari B (2005) Poverty, property rights and collective action: Understanding the distributive aspects of common property resource management. *Environ Dev Econ* 10:7–31.

Andrade GSM, Rhodes JR (2012) Protected Areas and Local Communities: an Inevitable Partnership toward Successful Conservation Strategies? *Ecol Soc* 17:art14.

Anthony B (2007) The dual nature of parks: Attitudes of neighbouring communities towards Kruger National Park, South Africa. *Environ Conserv* 34:236–245.

Appiah M, Blay D, Damnyag L, et al. (2009) Dependence on forest resources and tropical deforestation in Ghana. *Environ Dev Sustain* 11:471–487.

Babulo B, Muys B, Nega F, et al. (2008) Household livelihood strategies and forest dependence in the highlands of Tigray, Northern Ethiopia. *Agric Syst* 98:147–155.

Bardhan P (2002) Decentralization of governance and development. *J Econ Perspect* 16:185–205.

Bardhan P, Ghatak M, Karaivanov A (2007) Wealth equality and collective action. *J Public Econ* 91:1843–1874.

Barrett CB, Travis AJ, Dasgupta P (2011) On biodiversity conservation and poverty traps. *Proc Natl Acad Sci* 108:13907–13912.

Bennett EL (2011) Another inconvenient truth: the failure of enforcement systems to save charismatic species. *Oryx* 45:476–479.

Bennett NJ, Dearden P (2014) Why local people do not support conservation: Community perceptions of marine protected area livelihood impacts, governance and management in Thailand. *Mar Policy* 44:107–116.

Bonner R (1993) *At the hand of man: peril and hope for Africa's wildlife.* Knopf Doubleday Publishing Group.

Brechin SR, Wilshusen PR, Fortwangler CL, West PC (2002) Beyond the Square Wheel: Toward a more comprehensive understanding of biodiversity conservation as social and political process. *Soc Nat Resour* 15:41–64.

Brown K (2002) Innovations for conservation and development. *Geogr J* 168:6–17.

Cárdenas JC (2000) How do groups solve local commons dilemmas? Lessons from experimental economics in the field. *Environ Dev Sustain* 2:305–322.

Cavalcanti C, Schläpfer F, Schmid B (2010) Public participation and willingness to cooperate in common-pool resource management: A field experiment with fishing communities in Brazil. *Ecol Econ* 69:613–622.

Cavendish W (1999) *Poverty, inequality and environmental resources: quantitative analysis of rural households.*

Charnley S, Poe MR (2007) Community forestry in theory and practice: Where are we now? *Annu Rev Anthropol* 36:301–336.

FAO (2015) Global forest resources assessment. *FAO*. Roma, Italy.

Ferraro PJ (2002) Direct Payments to Conserve Biodiversity. *Science* 298:1718–1719.

Gatiso, T.T Vollan, B. Vimal, R. & Kuehl, H. (2017). If Possible, Incentivize Individuals Not Groups: Evidence from Lab-in-the-Field Experiments on Forest Conservation in Rural Uganda. *Conserv. lett, 1-11.*

Gatiso TT (2017) Households' dependence on community forest and their contribution to participatory forest management: evidence from rural Ethiopia. *Environ Dev Sustain* 1–17.

Gatiso TT, Wossen T (2014) Forest dependence and income inequality in rural Ethiopia: evidence from Chilimo-Gaji community forest users. *Int J Sustain Dev World Ecol* 22:1–11.

Jodha NS (1986) Common property resources and rural poor in dry regions of India. *Econ Polit Wkly* 21:1169–1181.

Kapoor I (2001) Towards participatory environmental management? *J Environ Manage* 63:269–279.

Kramer R, Schaik C van, Johnson J (1997) *Last Stand: Protected areas and the defense of tropical biodiversity.* Oxford University Press, New York.

Kumar DS, Palanisami K (2009) An economic inquiry into collective action and household behaviour in watershed management. *Indian J Agric Econ* 64:108–122.

López-Feldman A, Taylor JE, Yúnez-Naude A (2011) Natural resource dependence in rural Mexico. *Investig Económica* 70:23–44.

Mamo G, Sjaastad E, Vedeld P (2007) Economic dependence on forest resources: A case from Dendi District, Ethiopia. *For Policy Econ* 9:916–927.

MEA MEA (2005) *Ecosystems and Human Well-Being Ecosystems and Human Well-Being.* 109.

Mohammed AJ, Inoue M (2013) Forest-dependent communities' livelihood in decentralized forest governance policy epoch: case study from West Shoa zone, Ethiopia. *J Nat Resour Policy Res* 5:49–66.

Narain U, Gupta S, van't Veld K (2008) Poverty and resource dependence in rural India. *Ecol Econ* 66:161–176.

Nepal SK, Weber KE (1995) Prospects for coexistence: wildlife and local people. *Ambio* 24:238–245.

Nuesiri EO (2015) *Representation in REDD: NGOs and Chiefs Privileged over Elected Local Government in Cross River State, Nigeria.*

Olken BA (2010) Direct democracy and local public goods: Evidence from a field experiment in Indonesia. *Am Polit Sci Rev* 104:243–267.

Ostmann A (1998) External control may destroy the commons. *Ration Soc* 10:103–122.

Ostrom E (1999) Self-governance and forest resources. Center for International Forestry Research (CIFOR).

Ostrom E (1990) Governing the commons: The evolution of institutions for collective action. Cambridge University Press.

Persha L, Agrawal A, Chhatre A (2011) Social and ecological synergy: Local rulemaking, forest livelihoods, and biodiversity conservation. *Science* (80-) 331:1606–1608.

Pouta E, Rekola M, Kuuluvainen J, et al. (2002) Willingness to pay in different policy-planning methods: Insights into respondents' decision-making processes. *Ecol Econ* 40:295–311.

Pretty J, Smith D (2004) Social capital in biodiversity conservation and management. *Conserv Biol* 18:631–638.

Rahut DB, Ali A, Behera B (2015) Household participation and effects of community forest management on income and poverty levels: Empirical evidence from Bhutan. *For Policy Econ* 61:20–29.

Salafsky N, Wollenberg E (2000) Linking livelihoods and conservation: A conceptual framework and scale for assessing the integration of human needs and biodiversity. *World Dev* 28:1421–1438.

Smith RJ, Muir RDJ, Walpole MJ, et al. (2003) Governance and the loss of biodiversity. *Nature* 426:67–70.

Spiteri A, Nepal SK (2006) Incentive-based conservation programs in developing countries: A review of some key issues and suggestions for improvements. *Environ Manage* 37:1–14.

Sunderlin WD, Belcher B, Santoso L, et al. (2005) Livelihoods, forests, and conservation in developing countries: An overview. *World Dev* 33:1383–1402.

Terborgh J (1999) *Requiem for Nature.* Washington, DC: Island Press/Shearwater Books.

Treves A, Wallace RB, Naughton-Treves L, Morales A (2006) Co-managing human–wildlife conflicts: A review. *Hum Dimens Wildl* 11:383–396.

Vedeld P, Angelsen A, Sjaastad E, Berg GK (2004) *Counting on the Environment: Forest Incomes and the Rural Poor.* The World Bank Environment Department, Washington.

Wade R (1989) *Village Republics: Economic conditions for collective action in South India.* San Francisco.

Wells M, Guggenheim S, Khan A, et al. (1999) *Investing in biodiversity: a review of Indonesia' integrated consevation and development projects.*

Wells MP, McShane TO (2004) Integrating protected area management with local needs and aspirations. *AMBIO A J Hum Environ* 33:513–519.

West PC, Brechin SR (1991) *Resident peoples and national parks : social dilemmas and strategies in international conservation.* University of Arizona Press, Tucson.

Wilshusen PR, Brechin SR, Fortwangler CL, West PC (2002) Reinventing a Square Wheel: Critique of a resurgent "Protection Paradigm" in international Biodiversity Conservation. *Soc Nat Resour* 15:17–40.

Winterbottom B, Eilu G (2006) *Uganda biodiversity and tropical forest assessment, final report.* 64.

Wunder S (2013) When payments for environmental services will work for conservation. *Conserv Lett* 6:230–237.

World B (2015a). Population growth (annual %). *Data retrieved from World Development Indicators* (http://data.worldbank.org/indicator /SP.POP.GROW).

World B (2015b). Population density (people per sq. km of land area). *Data retrieved from World Development* (http://data.worldbank.org /indicator/EN.POP.DNST).

In: Forest Conservation
Editor: Pedro V. Eisenlohr

ISBN: 978-1-53614-559-5
© 2019 Nova Science Publishers, Inc.

Chapter 8

LARGE DAMS IN THE AMAZON: DISCONNECTING THE SOCIAL AND NATURAL SYSTEM

*Elineide Eugênio Marques and Adriana Malvasio**
Federal University of Tocantins, Palmas, TO, Brazil

INTRODUCTION

The disconnection between social and the natural systems is carried out at several dimensions, which can be noticed from small-scale events, such as local nutrient dynamics, or even by broader phenomena, such as climatic changes that influence the dynamics of water in environments, changing the regime or intensity of rain, for example. However, increasing the production of energy and inputs to satisfy the demands imposed by the current development model has contributed to the disconnection between water, forest and people, modifying the socio-environmental relations and the fauna are featured elements in this process.

---

* Corresponding Author Email: malvasio@mail.uft.edu.br.

In this study this question is analyzed from the large hydroelectric dams, built in cascade (Barbosa 1999), in the Tocantins River (Brasil - MME - EPE 2012; Tundisi 2005) and its effects on the fauna of fish and turtles. These two groups of animals depend on the dynamics of the river and the riparian forest for the maintenance of their populations. Some species also have symbolic importance and participate in the diet of the traditional indigenous and non-indigenous populations of the region (Begossi and Braga 1992; Andrade 2008; Salera Jr. et al. 2006; Cintra et al. 2007; Salera Jr. et al. 2007; Gomes et al. 2010; Souza et al. 2016).

The analysis was based on the experience of the authors who accompanied the environmental changes associated with the dam projects in the Tocantins River for more than a decade. In addition, the information contained in the studies and documents for the licensing and monitoring of fish and turtle fauna in the reservoirs of the region were considered. Some species of turtles are very important to the riverine communities, such as Amazonian turtle (*Podocnemis expansa*) and tracajá turtle (*Podocnemis unifilis*).

These communities usually prefer the consumption of scaled fish like piaus (various species of the genus *Leporinus, Schizodon, Hemiodus* and others), cachorra-verdadeira (*Hydrolycus* spp.), pacus and caranhas (several species of the genus *Myleus, Mylossoma* and *Colossoma*), papa-terra (*Prochilodus nigricans*) and branquinhas (species of the Curimatidae family), among others. On the other hand, large-scale species, usually migratory, sustain most commercial fisheries (surubim, *Pseudoplatystoma* spp., jaú, *Zungaro zungaro*, barbado, *Pinirampus pirinampu* and others) in Amazonian rivers (Santos et al. 1984; Doria et al. 2012; Barthem et al. 2016).

CHRONOLOGY OF THE DAMS CONSTRUCTION ON THE TOCANTINS RIVER

The Tocantins River has seven large hydroelectric plants in operation, which are Tucuruí (1984), Serra da Mesa (1998), Lajeado (2001), Cana Brava (2002), Peixe Angical (2006), São Salvador (2008) and Estreito (2010). The last five dams started operating in the last decade, indicating how fast these projects are being carried out (Table 1, Figure 1).

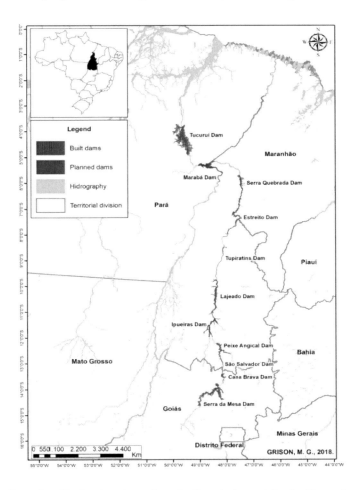

Figure 1. Cascade of hydroelectric projects built and planned for the Tocantins River.

Table 1. Hydroelectric plants in operation and planned for the Tocantins River

Name of the dam	State	Capacity (MW)	Reservoir area (km²)	Operational status	First year of operation
Serra da Mesa	Goiás	1275	1784	In operation	1998
Cana Brava	Goiás	450	139	In operation	2002
São Salvador	Tocantins	241	104	In operation	2009
Peixe Angical	Tocantins	453	294	In operation	2006
Ipueiras	Tocantins	480	933,5	Planned	-
Lajeado*	Tocantins	902	630	In operation	2001
Tupirantins	Tocantins	620	370	Planned	-
Estreito	Maranhão	1087	590	In operation	2010
Serra Quebrada	Maranhão	1328	420	Inventoried	-
Marabá	Pará	2160	1.115,4	Planned	-
Tururuí	Pará	8370	2850	In operation	1984

* or Luís Eduardo Magalhães Hydropower Plant.

The plan of the Electric Power Sector for the Tocantins basin (Brasil - MME - EPE 2012) also includes the projects of Ipueiras, Tupiratins, Serra Quebrada and Marabá, in different phases of licensing. In addition to these projects there are the Small Hydroelectric Power Plants (SHP) which are smaller and planned for the main tributaries. Tocantins state has more than 35 dams of this type planned for this basin.

IMPACTS OF DAMS ON THE FAUNA OF FISH AND TURTLES IN THE TOCANTINS RIVER

The flooding of large areas of riparian forest during the filling of hydroelectric reservoirs in the savannah region compromises plant biodiversity (Grison 2015), considering the great distinction that exists between phytophysiognomies in the area and modifying the dynamics of the aquatic environment and fish assemblages and turtles. The populations of several aquatic species depend on the riparian forest for their feeding and reproductive success.

The loss of feeding and spawning areas of fishes and turtles (flood plains, marginal vegetation, beaches) by the cascading dams built at Tocantins River contributed to the natural systems disconnection. A combination of flooded riparian vegetation and changes in flood regimes adversely affects populations that depend on the terrestrial and aquatic environments material in their diet.

Considering that the main species consumed by the communities in the region are predominantly herbivorous (fish – piaus and pacus - genus *Leporinus* and *Myleus*, for example; turtles species *P.expansa* and *P.unifilis*; Araújo-Lima et al. 1995, Pritchard and Trebbau 1984, Malvasio et al. 2003), the reduction of riparian vegetation and breeding sites affect directly the traditional communities. At the same time, the expansion of the environment with impoundment incorporates allochthonous resources in the aquatic system and favors opportunistic species. However, this condition is temporary (Hahn and Fugi 2007) and may vary in the different biotopes of the aquatic ecosystem (Araújo-Lima et al. 1995), which changes the quantity and quality of food available to the communities in the region. These problems were related by Araújo-Lima et al. (1995) for fish and Pezzuti et al. (2016), for turtles. The abundance of fish that eat fruits and seeds especially pacus (genus *Myleus, Mylesinus, Piaractus*), and piracanjubas (genus *Brycon*) tend to decline after the dam is built. The changes were noticed by fishers at Lajeado Dam downstream.

Reproduction in both groups is related to environmental factors such as precipitation, temperature and water level of rivers. The drought and flood regime are changed in the reservoirs, modifying the reproductive dynamics in these animals (Agostinho et al. 1999; Pezzuti et al. 2016; Lima et al. 2017).

Considering the reservoir regions, flooding may occur on the reproductive areas or can happen the permanent exposure of nesting sites, modifying, for example, granulometric aspects of the soil. In the case of the turtles, when changes occur in the sediments that form the nests, the sex ratio of the hatchlings can be altered as a function of interferences in the eggs incubation temperature. The grain size of the sediments is related to the soil temperature due to the size of the grains and their retention of heat. Other changes related to granulometry are focused on incubation time and hatching success (Ferreira Jr. 2003).

Reducing water velocity alters the structure of the freshwater environment through thermal and chemical stratification along the reservoir and increases the sedimentation rate. In this case, fish eggs and larvae drifting by the river did not appear in the samples collected in the lower half of the reservoir, indicating that the conditions are not suitable for incubation of the offspring. On the other hand, increasing water transparency due to sedimentation can elevate the larvae predation risk (Agostinho et al. 2007).

The fish fauna presents changes in the local species diversity, however the changes in the abundance of the populations are accentuated, mainly on long distance migratory species. The medium and large sizes sustain the fishing activity. Species such as caranha (*Colossoma brachypomus*) and surubim (*Pseudoplatysma fasciatum*) almost disappeared from middle Tocantins River stretch. Considering turtles, *P. expansa* stands out as a migratory species and it can move hundreds of kilometers to perform the posture, so the dams in series, as they have been occurring in the Tocantins river, can have serious implications in the environments occupied during the periods rainfall and drought, and interruption of gene flow is also possible (Pezzuti et al. 2016).

Other aspects that generally have a negative impact on fishes and turtles populations are increased boat flow and increased hunting/fishing pressure. The more intense traffic of boats, especially during the planning and implementation phases of the hydroelectric dams, causes disturbances in reproduction, and some species need to find other places to complete their reproductive cycle. The increase in the number of people, especially in the mentioned phases, predisposes these animals to a more hunting/fishing pressure.

At the same time, the distance between companies and communities observed during negotiation processes makes it difficult to implement conservation measures effectively. The local residents, who consume and/or trade these animals and observe their behavior daily, in a way, cooperate with their conservation.

The drastic changes observed in the areas of the dams, especially the mortality of fish and the capture of large numbers of individuals, also induce the change of the local inhabitants' behavior. Many people increase hunting and fishing pressure on stocks as a results of drastic and rapid changes in the region.

THE USE OF FISH AND TURTLES BY LOCAL COMMUNITIES

As already mentioned, many communities have an important relation with the mentioned species, considering the symbolic, cultural, economic and alimentary values, especially in riverside community and indigenous. Fishing is the most relevant economic activity in many Amazonian locations, in addition to fish being the main protein in the diet (Smith 1979a; Cerdeira et al. 1997).

The effects of hydroelectric dams change aquatic ecosystems and alter subsistence fishing. The installation of the reservoir requires modification of the way of life of the fishers and fisheries. The loss of lowland areas for the cultivation of subsistence crops, the environmental dynamics and the abundance of species requires the adaptation of fishing gear, vessels and fishing method (Gomes et al. 2010). Fish fauna structure is altered, allowing the proliferation of some species and the reduction or even elimination of others (Agostinho et al. 2008), which leads to a forced adaptation of the fishers.

The displacement forced by the hydroelectric building also causes suffering especially to the riverside residents, who saw their relationship with the river to disappear. The feeling of attachment to the place and symbolic losses reveal a deep impact on the lives of people who have lost touch with nature and its surroundings (Ertzogue et al. 2017).

In Brazil, turtles and fish of the Amazonian region have been used by man for food (meat and eggs) and the oil of eggs in the preparation of butter, in the lighting and manufacture of cosmetics (Cantarelli et al. 2014; Martins and Molina 2008; Smith 1979b). At the beginning of the twentieth century, butter ceased to be a commercial product, but the consumption of adult animals and their eggs remained, constituting until today a significant food resource for riverine populations, and important in the market of some cities (Andrade 2008, Fachín-Terán 2005, Pezutti 2003, Salera Jr. et al. 2007). In the festivals of these traditional communities, it is common to observe species of turtles and fish inserted in some form in the festive activities (mystical and alimentary), like typical dishes, special recipes for the magical beings; paintings made of annatto by the body, reminding these groups of fauna (Nascimento et al. 2015; Salera Jr et al. 2006; Tito 2013). Another use is in folk medicine. In Tocantins River, fish has been used to cure diseases as skin burns, rheumatis, parasites and others since early (Begossi and Braga 1992) and turtles are related with respiratory problems, considering the riverside community.

The relationship between damming of large rivers, changes in fish and turtle fauna, ecological and social characteristics of the areas have also been discussed, converging in a very controversial scenario and with actions still limited to avoid interference in the way of life of traditional communities. The change in the fauna´s composition is directly related to the interference in the way of life of the fishers, considering that the consumption of fish and turtles are important sources of protein input.

The small-scale fisheries make an important contribution to nutrition, food security, sustainable livelihoods and poverty alleviation, especially in developing countries (FAO 2007-2008). Despite this significant contribution, the issues constraining the sustainable development of small-scale fisheries remain poorly understood.

DISCONNECTING SOCIAL AND NATURAL SYSTEMS

The population decrease of many fish and turtle species can affect negatively the communities of the region that use this resource. It causes the detachment of social and natural systems in this case. At the same time, the deliberate distancing between companies that construct dams and local population hampers mitigation measures and aggravates socio-environmental issues in these areas particularly disconnecting natural and social systems.

Considering mitigating measures, it is clear that their establishment and even implementation, in many cases, would have to include the participation of communities that have the most interaction with these wildlife groups. In turtles, for example, one of the measures suggested in some cases would be a conservation management project, especially on the spawning areas, to protect the eggs and hatchlings until they went to the aquatic environment. The protection and recovery of riparian forests is another example, they are recommendations to protect the areas of reproduction of the fish and the integrity of the aquatic environment. In general, these types of programs also have Environmental Education actions. Of course, results in terms of conservation would be much more effective if people living with these species participate effectively in these actions (Cantarelli et al. 2014), which in many cases are indicated by researchers and environmental agencies, excluding the possibility working together.

The competent and adequate water governance requires as a priority water security for the human population and for ecosystem functioning (Tundisi 2017). These follows the "Human well-being and progress toward sustainable development are vitally dependent upon improving the management of Earth´s ecosystems to ensure their conservation and sustainable use. But while demands for ecosystem services such as food and clean water are growing, human actions are at the same time diminishing the capability of many ecosystems to meet these demands" (Alcamo et al. 2003, p. 1).

The fragmentation of information and teams that work in different areas within the same dam (intra-ventures), the disconnection between the dams (inter-ventures) and the discontinuity of information over time, with the overlapping of ventures in different phases in the same river (planning, licensing, monitoring) and the difficulty in showing to the communities the studies produced by the researchers, accentuate the disconnection between the social and natural systems.

Doria et al. (2017) analyzed the invisibility of fisheries and inadequacy of fishers' participation in process of hydropower development in the Amazon, focusing on gaps between legally mandated and actual outcomes, however that discussion needs to be enlarged.

The increase in the number of hydroelectric intensifies the disconnection processes and alters the original characteristics of the systems, as verified with the construction of cascading reservoirs in the Tocantins River. This scenario leads to greater damage to environmental issues and their conservation, not to mention countless impacts and consequences that are still unknown.

Considering these issues, the main challenges are related to an effective connection between the actors that participate in this process, as well as the investment in scientific research to aid in decision making. The interlinkages between social and natural systems have to become deeper in order to reduce the vulnerability of the human populations to environmental changes. How to predict or anticipate the disruption is a big challenge for the stakeholders in cascading reservoir rivers. Another challenge is to rethink the model of power generation in Brazil. In order to make progress, a real interaction is necessary, thinking about the conservation of natural resources as something holistic, without separating natural and social systems.

ACKNOWLEDGMENT

We thank the Federal University of Tocantins for the support received.

REFERENCES

Agostinho AA, Marques EE, Agostinho CS, Almeida DAD, Oliveira RJD, Melo JRBD (2007). Fish ladder of Lajeado Dam: migrations on one-way routes? *Neotropical Ichthyology*, 5(2): 121-130.

Agostinho AA, Miranda LE, Bini LM, Gomes LC., Thomaz SM, Suzuki HI (1999) Patterns of colonization in neotropical reservoirs, prognoses on aging. In Tundisi JG, Straskraba MS (ed), *Theoretical Reservoir Ecology and its application.* IIE—International Institute of Ecology, São Carlos: 227–265.

Alcamo J et al. (2003) *Ecosystems and Human Well-being: A Framework for Assessment (Summary)*, Island Press, Washington, pp. 1-266. http://pdf.wri.org/ecosystems_human_wellbeing.pdf

Andrade PCM (2008) Criação e Manejo dos Quelônios no Amazonas. [Creation and Management of Chelonians in the Amazon.] Editora Pro-Várzea/FAPEAM/SDS, Manaus.

Araújo Lima CARM, Agostinho AA, Fabré NN (1995). *Trophic aspects of fish communities in Brazilian rivers and reservoirs. Limnology in Brasil.* Rio de Janeiro: ABC/SBL 384p.

Barbosa FAR, Padisák J, Espíndola ELG, Borics G, Rocha O (1999) The Cascading Reservoir Continuum Concept (CRCC) and its Application to the River Tietê-basin, São Paulo State, Brazil. In Tundisi JG, Strakraba M (ed) *Theoretical Reservoir Ecology and its Applications.* São Carlos, IIE, BAS/Backhuys Publishers, pp 425-437.

Barthem R, Efrem F, Goulding M (2016) As migrações do jaraqui e do tambaqui no rio tapajós e suas relações com as usinas hidrelétricas. [The migrations of the jaraqui and tambaqui in the tapajós river and its relations with the hydroelectric plants.] In: Alarcon DF, Millikan B, Torres M (Orgs) *Ocekadi: Hidrelétricas, Conflitos Socioambientais e Resistência na Bacia do Tapajós.* Editora Câmara Brasileira do Livro, São Paulo, p 479-493.

Begossi A, Braga FMS (1992) *Food taboos and folk medicine among fishermen from the Tocantins River* (Brazil). Amazoniana. Kiel, 12(10): 101-118.

Brasil - MME - EPE. (2012) *Plano Decenal de Expansão de Energia 2021*, 1, pp 386.

Cantarelli VH, Malvasio A, Verdade LM (2014) Brazil's *Podocnemis expansa* Conservation Program: Retrospective and Future Direction. *Chelonian Conservation and Biology*, 13(1):124-128.

Cerdeira RGP, Ruffino ML, Isaac VJ (1997) Consumo de pescado e outros alimentos pela população ribeirinha do Lago Grande de Monte Alegre, PA-Brasil. *Acta Amazonica*, 27(3): 213-228.

Cintra IHA, Juras AA, Andrade JAC, Ogawa M (2007) Caracterização dos desembarques pesqueiros na área de influência da usina hidrelétrica de Tucuruí, estado do Pará, Brasil. *Bol Téc Cient CEPNOR*, 7:135-152.

Doria CRC, Ruffino ML, Hijazi NC, Cruz RL (2012) A pesca comercial na bacia do rio Madeira no estado de Rondônia, Amazônia brasileira. *Acta Amaz.*, 42(1):29-40.

Ertzogue MH, Ferreira DTAM, Marques EE (2017) "É a morte do Rio Tocantins, eu sinto isso": desterritorialização e perdas simbólicas em comunidades tradicionais atingidas pela hidrelétrica de Estreito ["It is the death of the Tocantins River, I feel it": deterritorialization and symbolic losses in traditional communities affected by the Estreito hydroelectric plant], TO. *Revista Sociedade & Natureza*, 29(1).

Fachín-Terán A (2005) Participação Comunitária na Preservação de Praias para Reprodução de Quelônios na Reserva de Desenvolvimento Sustentável Mamirauá, Amazonas, Brasil. [Community Participation in the Preservation of Beaches for Reproduction of Chelonia in the Mamirauá Sustainable Development Reserve, Amazonas, Brazil] *UAKARI*, Belém-PA, 1(1): 9-18.

FAO (2007-2018) Small-scale fisheries - Web Site. Small-scale fisheries. FI Institutional Websites. In: *FAO Fisheries and Aquaculture* Department. Rome. http://www.fao.org/fishery/.

Ferreira Jr PD (2003) *Influência dos Processos Sedimentológicos e Geomorfológicos na Escolha das Áreas de Nidificação de Podocnemis expansa (Tartaruga-da-Amazônia) e Podocnemis unifilis (Tracajá), na Bacia do Rio Araguaia. [Influence of sedimentological and geomorphological processes on the selection of nesting areas of Podocnemis expansa (Amazonian Turtle) and Podocnemis unifilis (Tracajá), in the Araguaia River Basin.]* Tese. Universidade Federal de Ouro Preto, Ouro Preto- MG.

Gomes KD, Marques EE, Parente TG (2010) *Percepções dos pescadores sobre as alterações ambientais e da pesca a jusante da barragem da usina hidroelétrica do lajeado*, [*Perceptions of fishermen on environmental and fisheries changes downstream of the dam of the slab hydroelectric plant*] Brasil, 10(1): 158-183p http://www.periodicos.rc. biblioteca. unesp.br/index.php/olam/article/view/3891.

Grison MG (2015) *Efeito da formação do reservatório da usina do lajeado no Rio Tocantins sobre a vegetal ripária. [Effect of the formation of the reservoir of the slab plant on the Tocantins River on the riparian plant.]* Dissertação. Universidade Federal do Tocantins.

Hahn NS, Fugi R (2007) A alimentação de peixes em reservatórios brasileiros: alterações e consequências nos estágios iniciais do represamento. [Feeding of fish in Brazilian reservoirs: changes and consequences in the initial stages of damming.] *Oecologia Brasiliensis*, 11(4): 469-480.

Lima MAL, Kaplan DA, Doria CRC (2017) Hydrological controls of fisheries production in a major Amazonian tributary. *Ecohydrology.* 2017;10:e1899. https://doi.org/10.1002/eco.1899.

Malvasio A, Molina FB, Sampaio, FAA (2003) Comportamento e preferência alimentar em *Podocnemis expansa* (Schweigger), *P. unifilis* (Troschel) e *P. sextuberculata* (Cornalia) em cativeiro (Testudines, Pelomedusidae). *Revista Brasileira de Zoologia,* 20 (1): 161-168.

Martins M, Molina FB (2008) Panorama geral dos répteis ameaçados do Brasil. [Overview of endangered reptiles in Brazil.] In: Machado ABM et al., (ed), *Livro Vermelho da Fauna Brasileira Ameaçada de Extinção.* MMA e Fundação Biodiversitas, Brasília e Belo Horizonte-MG, p 327-334.

Nascimento HS, Athayde SF, Lima MA (2015) Monitoramento Participativo da Pesca e Caça na Terra Indígena Paquiçamba, Volta Grande do Xingu, PA. *Relatório Técnico: Síntese e Análise dos Resultados do Primeiro Ano de Monitoramento, 2014. 2015.* Verthic Consultoria e Participações, Altamira.

Pezzuti, JCB, Vidal MD, Félix-Silva D (2016) Impactos da Construção de Usinas Hidrelétricas sobre Quelônios Aquáticos Amazônicos. In: Alarcon DF, Millikan B,Torres M, (org) Ocekadi: *Hidrelétricas,*

Conflitos Socioambientais e Resistência na Bacia do Tapajós. Editora Câmara Brasileira do Livro, São Paulo.

Pritchard PCH, Trebbau P (1984) *The Turtles of Venezuela*. Society for the Study of Amphibians and Reptiles, 1.

Salera Jr G, Franklim WG, Malvasio A, Giraldin O (2007) *Caça e Pesca entre os Índios Karajá do Norte, Terra Indígena Xambioá, Estado do Tocantins, Brasil*. Publicações Avulsas do Instituto Pau Brasil de História Natural, 10: 85-88.

Salera Jr G, Malvasio A, Giraldin O (2006) Relações Cordiais. [Cordial Relationships.] *Ciência Hoje*, 38: 61-63.

Santos G, Jegu M, Merona B (1984) *Catálogo de peixes comerciais do baixo Rio Tocantins. [Commercial fish catalog of the lower Tocantins River.]* Manaus: Eletronorte/CNPq/Inpa

Smith, NJH (1979a) A pesca no rio Amazonas. CNPq/Inpa.

Smith, NJH (1979b) Quelônios Aquáticos da Amazônia: um recurso ameaçado. [Amazonian aquatic chelonians: a threatened resource.] *Acta Amazônica Manaus*, 9(1): 87-97.

Souza MF, Marques EE, Miranda EB, Araújo AF (2017). Do rio Tocantins a Hidrelétrica de Peixe Angical: os peixes e as pescarias na memória dos pescadores. [From the Tocantins River to the Peixe Angical Hydroelectric Plant: fish and fisheries in the memory of fishermen.] *Revista Interface (Porto Nacional),* 12:119-134.

Tito MCA (2013) *Tinguizada Xerente: comida, conhecimento e cosmologia. [Tinguizada Manager: food, knowledge and cosmology*.] Dissertação. Universidade Federal do Tocantins.

Tundisi J G (2005) Gerenciamento integrado de bacias hidrográficas e reservatórios: estudo de caso e perspectivas. [Integrated management of river basins and reservoirs: a case study and perspectives.] In: Nogueira MG, Henry R, Jorcin A (ed.) (2005) *Ecologia de reservatórios: impactos potenciais, ações de manejo e sistemas em cascata*. São Carlos: Rima, pp 1-21.

Tundisi JG (2007) The exploitation of the hydroelectric potential of the Amazon region. *Estudos Avançados*, 59:109-117.

In: Forest Conservation
Editor: Pedro V. Eisenlohr

ISBN: 978-1-53614-559-5
© 2019 Nova Science Publishers, Inc.

Chapter 9

SELECTION AND PROPAGATION OF NATIVE TREE SPECIES FOR IMPROVING ECOLOGICAL RESTORATION

Sebastián Pablo Galarco[1], Maite Romero Alves[1], Patricia Boeri[2], Luciano Roussy[1], Marina Adema[1], Blanca Villarreal[1], María Valentina Briones[3], María de los Ángeles Basiglio Cordal[4], Tatiana Cinquetti[1], Diego Iván Ramilo[1] and Sandra Elizabeth Sharry[1,2,5,*]

[1]Wood Research Laboratory,
School of Agriculture and Forestry Sciences,
National University of La Plata, Buenos Aires, Argentina
[2]Integrated Unit for the Innovation of the Agri-Food System of Northern Patagonia, National University of Rio Negro,
Río Negro, Argentina

---

* Corresponding Author Email: ssharry@gmail.com.

[3]Biodiversity and Biotechnology Research Institute,
National Council of Scientific and Technical Research,
Buenos Aires, Argentina
[4]Ecological Coordination of Metropolitan Area - State society,
In vitro culture laboratory, Buenos Aires, Argentina
[5]Scientific Research Commission of Buenos Aires,
Buenos Aires, Argentina

INTRODUCTION

Forests play key roles in the water cycle, soil conservation, carbon sequestration, and habitat protection, including for pollinators. Forest ecosystem functions support the provision of ecosystem services to humans. These constitute the direct and indirect contributions of forest ecosystems to human wellbeing. In this context, ecosystem functions are a subset of the interactions between the ecosystem structure and the processes that underpin the capacity of an ecosystem to provide goods and services. It is essential to know the structure and functions of these ecosystems to be able to manage them in a sustainable way. In addition, their sustainable management is crucial for sustainable agriculture and food security (FAO 2016). For instance, forest disturbances are foreseen to increase (forest fires, invasive pests); and competing socio-economic demands for forest goods and services can result in multiple drivers of forest change (http://forest.jrc.ec.europa.eu/activities/forest-ecosystem-services/).
Simultaneously with the demand for forest products, there is a great need to conserve forest ecosystems, for both their ecological and aesthetic values. Efforts around the world are aimed at achieving sustainable forest management, an approach that balances economic, environmental and social objectives (FAO 2001). Forests and biodiversity are strongly interrelated. Biodiversity depends to a large extent on the integrity, health and vitality of forested areas. On the other hand, a decrease in forest biodiversity will lead to losses in forest productivity and sustainability. Therefore, sustainable forest management is oriented to support the provision of forest goods and

Selection and Propagation of Native Tree Species... 241

services, and to enhance biodiversity levels. Changes in the species richness and composition of an ecosystem may lead to parallel changes in the amount or quality of services provided by that ecosystem, including carbon sequestration pollination, or pest control that are indicative of a linear relationship between biodiversity and ecosystem services (Louman et al. 2010).

Among all life forms, trees require special attention for conservation for their myriad of ecosystem services, their high-level threat of extinction, and their socio-economic and cultural value (Oldfield et al. 1998; Dawson et al. 2014). However, for the conservation of trees to be successful there is a considerable number of challenges to overcome (Pritchard et al. 2014). To effectively using native trees species for restoration and as timber, food or active principle sources, we must know their characteristics, as well as their proper collection and propagation, culture and planting systems.

Globally, mixtures of native tree species are increasingly used to restore disturbed and degraded areas. However, one important factor limiting native species selection and diversification is the availability of quality planting materials. Restoration with native species requires the identification of tree species with readily available seed and mature propagation technology that is suitable to the local context. Propagation in nurseries is still an important consideration for the forest native species in Argentina. The maximum diversity method uses a large number of species, but as a consequence this approach is highly dependent on the capacity of local nurseries to propagate a large number of species (Knowles and Parrota 1995). As for large-scale restoration projects, a lack of high-quality tree germplasm from local nurseries is a major constraint to scale up the use of native species (Broadhurst et al. 2008; He et al. 2012). Thus, it is essential that nursery research expands our basic knowledge of native species propagation and selects species suitable for afforestation (Lu et al. 2016). Plants for restoration are normally obtained by seeds, cuttings, or seedlings (included vitro plants). Plant propagules (seeds, seedlings, and cuttings) for afforestation should come from the same local area as the restoration site to ensure the maintenance of local gene pools and the use of locally evolved ecotypes. In other words, species, subspecies, and varieties appropriate to a specific restoration program are used in order to restore the genetic health of a local community of native species. The use of locally grown plant materials can help prevent the spread of highly destructive fungal, or viral pathogens, as well as insect pests. Restoration should be part of the solution to forest loss, not part of the problem by inadvertently transmitting plant diseases and insects from place to place (Sailer 2006)

In recent years, interest in the propagation of native plants has been growing. Propagating native plants is a lengthy process, but an integral part of forest ecosystems reclamation efforts. Propagation from seed is the most commonly used method of propagating native trees and shrubs. It is preferable to propagating from cuttings, as propagation by seed produces greater genetic diversity. However, some desirable and ecologically important species are either difficult to propagate by seeds or time consuming. Thus, it is extremely important to develop researches about how to propagate a species of interest by vegetative propagation (macro and micro propagation). This can be achieved by combining classic propagation techniques with an understanding of the ecological and reproductive characteristics of the species. If we investigate how species perpetuates under natural conditions, we may be able to vegetative propagate the species and produce nursery stock in situations in which there are constraints on using seed propagation. Vegetative propagation is commonly found with species that have short seed life, low seed viability, or complex or delayed seed dormancy strategies. Because vegetative propagation is more expensive than seedling growing, the production system must be efficient.

Like other wood species, trees from the Argentinian forests are difficult to propagate by seed or vegetative methods. Biotechnological methods such as *in vitro* culture have been developed in the last few years as an alternative to conventional vegetative propagation. Biotechnology plays a very important role not only in the propagation of selected individuals (being used at a commercial level), but also in its short-term preservation, and it offers the possibility of preserving the propagated material in a medium-term period (cold storage) or long-term period, using cryopreservation. Biotechnology offers new tools for complementing classical forest tree propagation methodologies in order to manage Forest Genetic Resources

Selection and Propagation of Native Tree Species... 243

(FGR). The impressive achievements of the techniques of molecular biology and plant tissue culture in the last decades are in the background of the development of fields such as DNA markers, tree genomics, genetic transformation, cryopreservation and plant regeneration (expression of cellular totipotency) (Toribio and Celestino 2000). Biotechnology is important in the domestication of native species and plantation forestry as it helps in rapid multiplication of seedlings, replanting of degraded natural forests, development of tree disease diagnostic tools, conservation based on population genetics, and in the selection and breeding of new tree species. Biotechnology is also useful in the characterization of genetic diversity, conservation and in forestry health.

Particularly suitable for application to genetic resources is the micropropagation, useful for rapid multiplication. It is achieved through various approaches including organogenesis, somatic embryogenesis, cell culture, micro-cuttings, and embryo rescue between others. Micropropagation techniques have been tested in many tree species, notably *Pinus, Picea, Eucalyptus*, *Acacia* and *Populus*, among others. *In vitro* techniques have a clear role within *ex situ* conservation strategies for genetic resources, particularly where it is important to conserve specific genotypes or where normal propagules such as recalcitrant seed may not be suitable for long-term storage. These strategies involve the use of conventional micropropagation, restricted growth techniques and cryopreservation (Blakesley et al. 1996).

In Argentina indigenous forest tree species provide goods and services such as firewood, fruits, timber, poles, fodder, amenities, and environmental protection. The native forests of Buenos Aires province (Argentina) are strictly confined to the coastal strip of Río de la Plata (locally Monte Blanco forests and Talares forests, Figure 1) and to the Western region (locally Caldenal). Woody species found in the Talares and the Monte Blanco forest include *Celtis ehrenenbergiana* Gill. ex Planch (Tala), *Scutia buxifolia* Reissek (Coronillo), *Jodina rhombifolia* (Hook. & Arn.) Reissek (Sombra de toro), *Schinus longifolius* (Lindl.) Speg. (Molle), *Erythrina crista galli* L. (Ceibo), *Sesbania punicea* (Cav.) Benth. (Acacia mansa), *Phytolacca tetramera* Hauman (Ombusillo), *Parkinsonia aculeata* L. (Cina-cina), *Salix humboldtiana* Willd. (Sauce criollo), *Citharexylum montevidensis* (Spr.) Mold. (Espina de bañado), *Terminalia australis* Cambess. and *Acacia caven* (Molina) Molina, among others. These species have different traditions (medicinal, food, etc.) and uses in industry. There is lack of information on the biology of these native tree species, the different purposes they could be used for and the genetic variation between and within them. This lack of knowledge could lead to irreversible loss of genetic diversity even before any study on variability and potential use is made. The implementation of *in situ* and *ex situ* conservation strategies is considered absolutely necessary (Rivas et al. 2004).

At the Facultad de Ciencias Agrarias y Forestales de la Universidad Nacional de La Plata (FCAyF), we have established a germplasm bank and research in native species from Talares and Monte Blanco forests using several techniques, including biotechnology. At the Laboratory of Research on Wood (LIMAD) several development projects are being run by different investigators, which include afforestation or restoration component, particularly in forest ecosystems near our Institution location in La Plata. As a result, we have been researching and implementing propagation systems of native trees over the last 3 decades. The final objective is to deliver plants into the ecosystem. Besides, we are doing an effort to share information with nursery growers and plant propagators. Native plant cultivation is a tool to promote biodiversity conservation and to enhance population awareness and interest in natural ecosystems.

The purpose of this chapter is to present new approaches to characterization, propagation and conservation of native tree species from Buenos Aires Talares and Monte Blanco forests. These results will be described based on the major methods being used, developed and applied to propagate and conserve native species.

ARGENTINIAN FORESTS SITUATION

Argentina is situated in the Southern cone of the American continent. It covers 3.761.274 km^2 and hosts approximately 33 million hectares of native forests. In 1914, Argentina was estimated to have more than 106 million hectares of native forests; by 1996, when a national action programme against desertification began, only 36 million hectares remained. Argentina, often perceived as a vast fertile territory, is losing its native forests. Today, the country's forests are vanishing at a rate of more than 829,000 hectares per year, mainly where agriculture is pushing into native forests (Belluscio 2009). Four different climates and associated forest formations can be identified: temperate mountains of the southern Andes, the Chaco formation on the border with Paraguay and Bolivia (a typically semi-arid subtropical zone), the so-called pampas, a very flat and treeless zone in the central part of the country where most of the big cattle ranches are located, and the Patagonia zone with its desolate steppes and poor soils (FRA 2000).

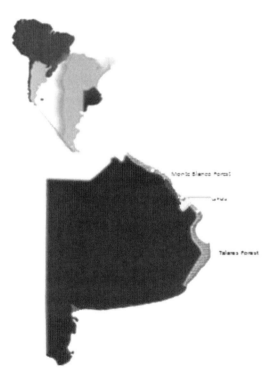

Figure 1. Location of natural forests (Talares and Monte Blanco) in Buenos Aires Province, Argentina.

Argentina is facing a true forest emergency, enhanced in the last 15 years by an uncontrolled expansion of the agricultural frontier. Historically, native forests have not been properly regulated, and the first federal law to regulate logging is the National Law 26.331 of Minimum Standards for the Environmental Protection of Native Forests, locally "Ley de Bosques" (Belluscio 2009). In 2007, the enactment of the law of native forests determined a barrier to the advance of soybean cultivation, with a remarkable support from the civil society. The law was a response to the environmental problems and social conflicts that have accompanied the high rate of deforestation over many past decades, due to a combination of new agricultural techniques, increased rainfall range and sustained international demand for grains and animal proteins that has led investors and producers out of the country's traditional agricultural region (the Pampas) and into the less favourable soils and climates of the northern provinces (Luft 2013). The law states that each province is required to start a participatory Native Woodland Use Planning Process (OTBN in Spanish) for their native forests.

BUENOS AIRES FORESTS

Buenos Aires Province covers 308.000 km², at average latitude of 37-38° SL, in the neotropical region, and hosts about 1500 native plant species. The number of extinct species due to overgrazing, intensive agriculture or urban settlements is not known. Thirty six percent of the Buenos Aires province area is used for agricultural purposes, 53% for livestock rising, while 11% is not being used, although it practically does not harbour any remaining pristine forests. Buenos Aires has recently (December of 2016) concluded their OTBN (territories of native forest management) official project. It was the last province to agree on the areas to conserve under the new Forests Law 26.331.

The native forests of Buenos Aires province, located in the southern cone of South America, are strictly confined to the coastal strip of Río de la Plata (coastal forests, locally "Talares" and "Monte Blanco") and to the western region (locally Caldén forests) (Parodi 1940). When the city of

Buenos Aires was founded, in 1590, its surrounding areas hosted forests of *Celtis ehrenenbergiana* (ex *Celtis tala*) and white algarrobo (*Prosopis alba*) (Cozzo 1992). During the colonization period of this territory, indigenous traditions and culture became less prominent as the colonial perceptions became more dominant. From the actions of the colonialists, it is evident that they considered forests to be unproductive lands in contrast to agriculture and ranching (Gabay et al. 2011). The use of forests for firewood was so intensive that the resources were exhausted at the beginning of the 19th century. Currently there are only remnants or relict of these forests locally called "Talares," "Caldenal" and "Monte Blanco," as mentioned before. The native dry forest dominated by *Celtis erhenbergiana* (Ulmaceae) and *Scutia buxifolia* (Rhamnaceae) constitutes the main woodland community of the eastern plain called Pampa in Buenos Aires province (Parodi 1940; Cabrera 1976). This woodland has undergone an important process of degradation due to the increasing urban and agricultural areas as well as the use of wood for fuel. Besides their historical value, an urgent call is made for their preservation, sustainable use and defense from biological invasions (Lewis et al. 2009). The Talares forests area is an environmental mosaic where thorny forests and woodland patches combine with humid grasslands and coastal communities of Río de La Plata. It is a xerophyte forest type structured by few tree species surrounded by a lower and moister soil matrix. The forest grows on highly calcareous parent materials, derived from sea transgressions and regressions in the Quaternary (Cabello and Arambarri 2002). The current status of conservation shows that the protection of these forests is considered a priority due to their high biodiversity and fragility. The unique characteristics of the Talares led to the creation, in 1984, of the Biosphere Reserve "Parque Costero del Sur" (MAB-UNESCO), now considered a natural and cultural heritage site. This reserve consists of a group of natural interphases, with exceptionally rich flora and fauna. Forests dominated by *Celtis ehrenbergiana* are becoming susceptible to invasion by exotic species as *Ligustrum lucidum* (privet). The degree of invasion has exceeded the threshold of irreversibility, consolidating as neo-ecosystems dominated by privet (Diaz Villa et al. 2016). The tala tree is usually common in the riparian zone, although human activities limit its propagation. It is a valuable species for the multiplication of native animals as it serves as shelter, food source and bird nesting. Ribichich and Potomastro (1998) found that the dominant trees are *Scutia buxifolia* and *Celtis ehrenbergiana*. There is also *Schinus polygamus*, which presents the bell-shaped structure of the pioneer species. Over 1200 m distant from the river, this species is co-dominated by *S. buxifolia* and by *C. ehrenbergiana* trees regenerated from stumps. The differences between the old-growth stands seem to be related to the gradients of soil texture and nutrient concentration, raising edaphic stress towards the river. The stress tolerance of *S. buxifolia*, and the aptitude of *S. polygamus* to recruit in disturbed habitats seem to have prevented the post-logging recruitment of *C. ehrenbergiana*. Tala regenerated possibly due to a better competitive performance in a more favourable site. Ribichich and Potomastro (1998) recommend the restoration of the qualitative features and the control of privet. Goya et al. (1992) and Arturi and Goya (2004) have conducted several studies in order to protect the species. The reproductive system varies among the different species: *C. ehrenbergiana* is anemophilous and selfcompatible. *S. buxifolia* is entomophilous and floral visitor's dependant. *J. rhombifolia* is entomophylous, although spontaneous autogamy could favour reproduction in the absence of pollinators. S. *longifolia* could be an ambophilous species (pollinated both by insects and by the wind). This dual system may be the result of a system flexibility mechanism or an evolutionary transition (Torreta and Basilio 2013).

Other relict forest ecosystem in Buenos Aires is the locally named Monte Blanco. This forest, that originally occupied the elevated border areas of the Low Delta islands of the Río Paraná and some areas of Río de La Plata river coast, has almost been eliminated as a consequence of productive activities carried out in the region during the past century. At present, only relict patches with scarce regional representation may be found along the northeastern part of the Río de La Plata shore. Many of those forests have been abandoned, resulting in secondary forest formation that is subject to numerous invasive exotic species, such as *Ligustrum lucidum, Morus alba, Rubus ulmifolius, Gleditsia triacanthos, Fraxinus pennsylvanica* and other species. A priori observations suggest that successional trends do not lead to recovery of the original forest. Because of the low density of native tree species, it is difficult to predict their future persistence. In conclusion, ecological restoration strategies will be needed in order to increase native tree species richness and forest biodiversity in the Lower Delta of the Paraná River, the original forest, locally referred to as the Monte Blanco forest (Kalesnik et al. 2008).

METHODS FOR CONSERVATION AND PROPAGATION OF NATIVE FOREST SPECIES FROM TALARES AND MONTE BLANCO

At the FCAyF we proposed the creation of a Germplasm Bank for the conservation of native forest genetic resources from Talares and Monte Blanco. Our work in plant propagation started in 1983, mainly with forestry species, through conventional methods (macro propagation, seed technology) and biotechnology (Guido et al. 1999; Sharry et al. 2011; Sharry and Abedini 2014). Since 1999, data on target species have been collected in several natural areas, including the restricted Otamendi Natural Reserve, the provincial Pereyra Iraola Park, the Punta Lara reserve and the "Parque Costero del Sur" Biosphere Reserve (Rivas et al. 2004), following a methodology proposed by Painting and others (1993). Trees and seed-producing stands have been selected and marked, while material has been collected for the herbarium. Most of the samplings are used to develop the propagation protocols and analyse the seeds. The project included studies of the following species, among others: *Acacia caven* (Espinillo) *Parkinsonia aculeata* L. (Cina-cina) *Erythrina crista* galli L. (Ceibo), *Celtis ehrenbergiana* (Tala), *Scutia buxifolia* (Coronillo), *Citharexylum montevidense* (Espina de bañado), *Terminalia australis* (Palo amarillo), *Salix humboldtiana* (South American native willow), *Phyllantus sellowianus* (Sarandi blanco), *Jodinia rhombifoli*a (Sombra de toro) and *Phytolacca tetramera* (Ombusillo) among others. The following multiplication techniques were used: macro propagation (rooting of branches, roots, grafts) and micro propagation (by organogenesis, micrografts, somatic embryogenesis, embryos rescue). Complete plants have been obtained using these techniques. Individuals are conserved *ex situ* in the bank and several of them were delivered into the ecosystem. Seeds are being collected for conservation, after physiology studies have been carried out (feasibility and germination tests) according to FAO/IPGRI gene bank methodologies. The facilities and the experienced personnel allow the FCAyF to operate as a Centre of Certifying of Forestry Species Seeds, thus facilitating seed exchange with other institutions (Rivas et al. 2004). Botanical descriptors were conducted and the protocols of germination of seeds of acacia and tala parameters were adjusted. In addition, woody indigenous species including tala, espinillo and cina-cina originating from Pereyra Iraola and Punta Lara Parks have been planted in the Ecological Park of La Plata Municipality (Buenos Aires) with the aim of evaluating growth and behaviour.

SUCCESS IN PROPAGATION SYSTEMS

Below we present a brief description of the main forest species from Talares and Monte Blanco, which were successfully propagated and led to propagation conditions in nursery.

Celtis ehrenbergiana ("tala")

It is a deciduous, multi-stemmed, middle sized tree, 4 - 12 m tall and up to 80 cm in d.b.h. It has extended and heavily branched crown-wearing zigzag shaped twigs with paired thorns (Lahitte and Hurrell 1994). This tree exhibited phenotypic plasticity. This means it has advantages over other less adaptable species and might survive to climate changes (Nughes et al. 2013). Reproduction of the species in its natural range is by seed but high mortality rates occur during germination and seedling stages (Arturi 1997), resulting in a significant constraint for the conservation and management of the

Selection and Propagation of Native Tree Species... 251

"Talares." In the Centre of Certification of forest seeds, standards germination parameters were adjusted based on International Rules for Seed Testing (ISTA). Ramilo and Abedini (2007) achieved cuttings rooting using 50 ppm NAA. In order to do that, they collected lignified juvenile stem cuttings in winter and treated them with 4 different concentrations of IBA and NAA (0 (control); 25; 50; 75 and 100 ppm) for 48 h in the dark. The cuttings were planted in 90 cm^3 cell capacity block containers filled with perlite-vermiculite (1:1) substrate. After 5 months, higher survival was recorded for NAA, 71% treatment (50 ppm) versus a mean survival of 2,5% for IBA. Control (0 ppm) and low concentrations levels formed callus. On the other hand, we tried unsuccessfully to adjust the micropropagation, from the thorns, leaves and nodal sections. All explants form calluses without sprouts (Abedini and Ruscitti 1996; Adema et al. 2007). This species is very recalcitrant, like *Scutia*.

Salix humboldtiana ("South American Native Willow")

In South America, there is only one native species of willows, *Salix humboldtiana*, which belongs to the Salicaceae family. The leaves are lanceolated and finely toothed. It is the only willow that conserves the same colour on both sides of the leaves. In Argentina it extends from the north of the country to the province of Chubut in the south (Lahitte and Hurrell 1994). We adjust the macro and micropropagation of this species (Abedini et al. 2008; Adema et al. 2010; 2014). Macro propagation was achieved by rooting of cuttings (30 cm. long and 0.8 - 1.5 cm in diameter). They were immersed in 50 mg. L-1 IBA during 24 hours. The substrate was a mixture of earth-perlite-vermiculite (6:3, 5:0, 5). On the other hand, micro propagation is possible form nodal sections, cultured on Woody Plant Medium (WPM, Lloyd and McCown 1981) without growth regulators. *In vitro* rooting occurs on WPM supplemented with 0.1 mg. L-1 of IBA.

Parkinsonia aculeata ("Cina-cina")

Belonging to the Legumes family, Cesalpinoideae subfamily, it is a tree up 3 to 6 meters high with a 0,30 meters of trunk diameter. Its green and young branches turn into a dark bark when it grows. Its wood is fairly hard and fairly heavy (Tinto 1977) and it is used for firewood, textile use and paper industries (Lahitte and Hurrel 1994). It is cultivated as an ornamental and melliferous species (Novara 1984). It multiplies by seeds and it is capable to produce epicormic buds (Lahitte and Hurrel 1994). For *in vitro* germination, seeds were disinfected with a 50% of commercial sodium hypochlorite for an hour; after that, the seeds were treated with hydrogen peroxide (5% v/v).

This procedure was made with the purpose of breaking the dormancy of the seeds and to favour the disinfection of tegument. The *in vitro* germination of the seeds started 15 days before sowing in MS culture medium. The micropropagation was achieved from nodal segments in Broadleaved Tree Medium (BTM, Chalupa 1983), added with 9.29 µM Kinetin (KIN). The explants differentiated shoots and roots (Abedini 2005). The macropropagation can be obtained from juvenile cuttings collected during springtime. Rooting occurred using either 5 ppm of NAA solution or with 100 ppm of IBA solution at which rooting percentages reached 41% and 38%, respectively.

Erythrina crista-galli (*"Ceibo"*)

Belonging to the Legumes family, this native forest tree from Argentina, Uruguay, Brazil and Paraguay has ornamental, medicinal, ecological and industrial uses. It is also the national flower from Argentina. The *in vitro* protocol for Ceibo regeneration was adjusted from nodal sections of seedlings, which were obtained from seeds, germinated *in vitro*. Shoot growth was induced in Murashige and Skoog (1962) medium (MS) with

1mg.L-1 of BAP and 0.5 mg. L-1 of NAA. These shoots were subcultured in a WPM medium with 0.1 mg. L-1 of IBA where they elongated. The shoots formed roots in WPM medium with 0.1 mg. L-1 of NAA (Ruscitti and Abedini 1996; Abedini et al. 1997).

Terminalia australis ("Palo amarillo")

It is a hygrophilous-habit multi stemmed tree species that always lives near water. It naturally seeds itself. It bears fruit in summer and autumn. Popular medicine mentions the use of the infusion made with its bark as astringent due to the presence of tannins. Carpano et al. (2003) demonstrated the antifungal activity. Plants grow naturally in their habitat in isolation. For macropropagation we used cuttings (200 mm long and 8 mm in diameter on average) immersed in 100 ppm NAA solutions (Aquila et al. 2001).

Acacia caven (*"Espinillo"*)

It belongs to the family of Legumes, Mimosoideae subfamily. This tree is highly resistant to drought and has a big sprouting capacity. It is a little tree up to 5 meters high that has composed and deciduous leaves with two spinescent stipules as protection from animals that also turn the forest in a hard impenetrable one. The golden-yellowed flowers become a strong, gross and darkened legume fruit. We adjusted a micropropagation system through somatic embryogenesis. The explants were cotyledons obtained from mature seeds. The induction of somatic embryogenesis occurred in Murashige and Skoog (1962) medium (MS) at half strength concentration of macro and micronutrients with 1 mg.L-1 of 2,4-D and 0.1 mg/L-1 of BAP. The conversion of somatic embryos in a globular stage (30%) occurred on MS medium without PGR (Marinucci et al. 2003; Marinucci et al. 2005).

Phytolacca tetramera ("Ombusillo")

It is a dioecious species, belonging to the family Phytollacaceae. It is an endemic geophyte shrub of Buenos Aires. All species of this family are widely used in folk medicine for the treatment of different diseases (Gattuso et al. 1998; Zacchino 2000). The micro propagation is possible via organogenesis from nodal and internodal sections cultivated on MS culture medium, supplemented with 3% sucrose and IBA and 6- Bencil Amino Purine (BAP) at 0.5 ppm. After 60 days, whole plants can be transplanted under filed conditions (Basiglio Cordal and Sharry 2012; Basiglio Cordal et al. 2014).

Phyllantus sellowianus ("Sarandí blanco")

It is an Euphorbiaceae native shrub, 2-3 m. high, glabrous, with woody stems divided into branches from the base that grows near small streams and river-bank grasses. This native species is an ecologically important perennial woody shrub, which is decreasing rapidly as a result of environment degradation and indiscriminate collection. Rivas and Abedini (1996) adjusted the micropropagation through direct organogenesis induced from nodal and internodal segments. The best rooting of nodal and internodal micro cuttings was achieved in BTM basal medium, with 1.07 µM of NAA, in absence of cytokinins after three weeks, without basal callus formation. Rooted plants were successfully hardened in the greenhouse (Abedini and Rivas 1995).

Citharexylum montevidense ("Espina de bañado")

It is a native woody shrub used in handmade carpentry, building, poles and frameworks. Cuttings, which were 1 m. long and 3-15 mm. in diameter on average, were used for macropropagation. The experiments were carried out under environmental conditions inside a greenhouse with 25°C +/- 2°C. The thicker cuttings were put maintaining the same polarity in 25% perlite and 75% of black soil. The vegetative macropropagation is possible from cuttings 7 mm. in diameter and with material less than 1 year old after 110 days (Roussy et al. 2011).

Scutia buxifolia ("coronillo")

It is a small, thorny and slow-growing evergreen tree, 3 to 10 meters tall and up to 50 cm in d.b.h. that belongs to the Rhamnaceae family. Flowers are very small, green-yellowish, hermaphrodite and turn into a globose drupe fruit. Leaves are opposite, simple and elliptical, and branches are typically thorny. It produces reddish, fine-grained and high-density wood (specific gravity: 1,2) that is mainly used as firewood. It is a source of non-timber forest products: saponins are extracted from foliage and stems, fruits are locally utilized as animal fodder and roots are used in popular medicine. In nature, it propagates by seed, therefore seed availability is very important for management and conservation.

We are exploring vegetative propagation by means of rooting stem cuttings (Ramilo and Abedini, unpublished). Stem cuttings collected in winter were partially submerged for 24 hr period in 5 concentrations of IBA (0 ppm (check); 25 ppm; 50 ppm, 75 ppm, 100 ppm) and 5 of NAA (0ppm (check); 25 ppm; 50 ppm, 75 ppm, 100 ppm). Cuttings were planted in 90 cm^3 cell capacity block containers filled with perlite-vermiculite (1:1 ratio) substrate. After 5 months, 47% survival was observed with IBA versus 9% of NAA. The best overall survival and rooting response was achieved with 25 ppm IBA (96, 2% survival with longer roots per cutting). For NAA treatments, local check (0ppm) and 10 ppm showed the highest survival values (15, 4% and 11,5% respectively) but no correlation between survival and root length was observed.

CONCLUSION AND CHALLENGES

Targets have been set to restore 15% of degraded land by 2020 (CBD 2010) but restoration efforts are often constrained by lack of knowledge of the biology, propagation and management of tree species or by a lack of sufficient seed (Merritt and Dixon 2014). Consequently, restoration initiatives rarely incorporate less well-known species or those that are difficult to grow and to find in the market (Hoffman et al. 2015). We describe challenges associated with the restoration of threatened trees in the Talares and Monte Blanco forests of Northeast of Buenos Aires and analyse the effectiveness of methods used to define target species, identify seed sources and generate information on the biology of rare or threatened tree species. As mentioned above, the forest resources in the Northeaster region of Buenos Aires province are under significant threats and require urgent action to halt the perceived loss of forest genetic resources, so knowledge, characterization, propagation, and *ex situ* germplasm conservation programme of native forest species are fundamental. Despite the many challenges associated with collection of seeds and adjust propagation systems from rare and threatened native species, our data represent a significant first step towards improving the genetic diversity and species richness of the seedlings produced in the Buenos Aires forests. The methods described here may be of relevance to ecological restoration programmes elsewhere where the challenges of including rare and threatened tree species in germplasm bank, seedling production and planting operations are likely to be similar.

The currently objectives of the germplasm bank are to increase the number of accessions per species, increase the number of tree and shrub species, carry out molecular characterization of selected individuals, join forces with government agencies to develop *in situ* conservation schemes in protected areas, obtain additional reproductive material to establish field tests, exchange it with other scientists and make these species available to the native ecosystem restoration program. *Ex situ* strategies and propagation skills conserve unique species. These actions can lead to greater impact if supported by greater efforts to create new knowledge and more available collections to the scientific, nurseries, forestry and communities. Throughout this chapter we have used examples from the Buenos Aires relict forests, as a way of illustrating wider principles that can be applied in other countries. Future development of current research programs, the adherence to conservation policy and the expanding needs for education must be considered.

REFERENCES

Abedini W. and Rivas M.C. (1995) Obtención de plantas y callos *in vitro* de *Phyllanthus sellowianus* Mueller Arg. (Sarandí blanco). Actas IX Congreso Nacional de Recursos Naturales Aromáticos y Medicinales, Jujuy, Argentina. pp. 27. [In vitro plants and callus obtaining of *Phyllanthus sellowianus* Mueller Arg. (Sarandí blanco). National Congress of Aromatic and Medicinal Natural Resources. Jujuy, Argentina. pp. 27]

Abedini W., Boeri P., Galarco S., Huergo L., Lede S., Marinucci L., Rivas M.C., Ruscitti M., Sharry S. (1997) Vegetative propagation of native forest species in order to restore degraded ecosystems. *Proceedings International Symposium on Biotechnology of Tropical and Subtropical Species, Australia.*

Abedini W., Ruscitti M., Scelzo L. (1997) Propagación vegetativa de dos especies nativas de uso medicinal: *Celtis tala* (tala) y *Erythrina crista-galli* (Ceibo). *Proceedings VI Congreso Italo-Latinoamericano de Etnomedicina. Antigua Guatemala.* Guatemala. [Vegetative propagation of two native species for medicinal use: *Celtis tala* (tala) y *Erythrina crista-galli* (Ceibo). *Proceedings VI Congreso Italo-Latinoamericano de Etnomedicina. Antigua Guatemala.* Guatemala].

Abedini W., Boeri P., Marinucci L., Ruscitti M., Scelzo L. (2000) Biotécnicas aplicadas a especies forestales nativas. *Rev Inv Agr INIA Sistemas y Recursos Forestales* 9 (1):133-147. [Biotechniques applied to native forest species. *Rev Inv Agr INIA Sistemas y Recursos Forestales* 9 (1):133-147].

Abedini W. (2005) Propagación vegetativa de *Parkinsonia aculeata* L. por estaquillado. *Quebracho* 12:23-3. [Vegetative propagation of *Parkinsonia aculeata* by cuttings. *Quebracho* 12:23-3.]

Abedini W., Adema M., Herrera J., Sharry S., Villarreal B., Nikoloff N. (2008) Recursos Forestales nativos de la provincia de Buenos Aires: la biotecnología como una estrategia de conservación. Actas III. Congreso Nacional de Conservación de la Biodiversidad. Argentina. [Native Forest Resources of the province of Buenos Aires: biotechnology as a conservation strategy. Proceedings III National Congress of Conservation of Biodiversity].

Adema M., Besteiro S., Garcia Tartalo P., Sharry S., Abedini W. (2007) Propagación *in vitro* de *Celtis tala* Ex Planch. [In vitro propagation of *Celtis tala* Ex Planch.] *Actas XXII Jornadas Forestales de Entre Ríos*, Argentina.

Adema M., Basiglio Cordal M., Briones V., Villareal B., Ciocchini G., Abedini W., Sharry S. (2014) Macro y micropropagación de *Salix humboldtiana* y *Salix babilónica*. [Macro and micropropagation of *Salix humboldtiana* y *Salix babilónica*.] *Evaluación de la capacidad de remoción de Cu. Actas Jornadas de Salicáceas 2014- IV Congreso* Internacional de Salicáceas, Argentina..

Adema M., Curuchet G., Abedini W., Sharry S., Briones V., Villareal B., Basiglio Cordal MA., Ciocchini G. (2010) Estudio preliminar de la fitorremediación de cobre mediante *Salix humboldtiana*, [Preliminary study of copper phytoremediation using Salix humboldtiana] *Actas VII Encuentro Latinoamericano y del Caribe sobre Biotecnología Agropecuaria- REDBIO México 2010*, México.

Aquila S. V., Bonicelli V., Carparo S., Rossi J., Castro M. T., Spegazzini E., Abedini W. (2001) Propagación vegetativa de *Terminalia Australis* Cambes "Palo Amarillo," Combretaceae, *Actas IV Congreso Internacional de Plantas Medicinales, Chile.* [Vegetative Propagation of *Terminalia Australis* "Palo Amarillo," Combretaceae, Proceedings IV International Congress of Medicinal Plants].

Arturi M (1997) *Regeneración de Celtis tala Gill ex Planch en el NO de la Provincia de Buenos Aires* [*Regeneración de Celtis tala Gill ex Planch en el NO de la Provincia de Buenos Aires*] (Ph.D. thesis) Facultad de Ciencias Agrarias y Forestales, La Plata University. SEDICI Web http://sedici.unlp.edu.ar/handle/10915/4658.

Arturi M. F., Goya J. F. (2004) Estructura, dinámica y manejo de los talares del NE de Buenos Aires. [Structure, dynamics and management of the talares of the NE of Buenos Aires] In: Arturi M., Frangi J., Goya J. (eds.), *Ecología y Manejo de los bosques de Argentina*, EDULP, Argentina, 10:1-23.

Basiglio Cordal M., Sharry S. (2012) Conservación de especies medicinales de la Provincia de Buenos Aires, el caso del Ombusillo. [Conservation of medicinal species from the Province of Buenos Aires, the case of Ombusillo. Bioethics Notebooks.] *Cuadernos de Bioética*, Ed. Ad Hoc, Argentina 17: 45 - 47.

Basiglio Cordal M., Adema M., Briones V., Villarreal B., Panarisi M., Abedini W., Sharry S. (2014) Induction of somatic embryogenesis in *Phytolacca tetramera*, medicinal species of Argentina. *Emirates Journal of Food and Agriculture* 26(6): 552 - 557.

Belluscio A. (2009) Argentina's forests dwindle. Biodiversity at risk as forests give way to desert, *Nature International Weekly Journal of Science*, Published online doi:10.1038/news.2009.984. Accessed: 26 January 2018.

Blakesley D., Pask N., Henshaw G., Fay M. (1996) Biotechnology and the conservation of forest genetic resources: *in vitro* strategies and cryopreservation. *Plant Gr Reg* 20 (1): 11-16.

Broadhurst L., Lowe A., Coates D., Cunningham S., McDonald M., Vesk P.A., Yates C. (2008) Seed supply for broad scale restoration: maximizing evolutionary potential. *Evol Appl* 1: 587–597.

Cabello M., Arambarri A. (2002) Diversity in soil fungi from undisturbed and disturbed *Celtis tala* and *Scutia buxifolia* forests in the eastern Buenos Aires province (Argentina). *Microbiol Res* 157:115–125.

Cabrera A (1976) Regiones fitogeográficas argentinas. [Argentine phytogeographic regions.] *Enciclopedia Argentina de agricultura y jardinería.* ACME S.A.I.C. Buenos Aires.

Carpano S.M., Spegazzini E.D., Rossi J., Castro M., Debenedetti S. (2003) Antifungal Activity of *Terminalia australis*. *Fitoterapia* 74 (3): 294-297.

Convention of Biological Diversity (2010) https://www.cbd.int/ Accesed: 29 January 2018.

Cozzo D. (1992) Las pérdidas del primitivo paisaje de bosques, montes y arbustiformes de la Argentina con especial referencia a sus territorios áridos y húmedos. [Losses of the primitive forests landscape, mountains and shrubs of Argentina with special reference to its arid and humid territories.] *Acad. Nac. de Cs Miscelánea* 90.

Chalupa V. (1983) Micropropagation of Conifer and Broadleaved Forest Trees. Communicationes Instituti Forestalis. *Cechosloveniae* 13: 7-39.

Dawson I.K., Leakey R., Clement C., Weber J., Cornelius J., Roshetko J., Vinceti B., Kalinganire A., Masters E., Jamnadass R. (2014) The management of tree genetic resources and the livelihoods of rural communities in the tropics: non-timber forest products, smallholder agroforestry practices and tree commodity crops. *For. Ecol. Manage* 333: 9–21.

Díaz Villa M., Madanes N., Cristiano P., Goldstein G. (2016) Composición del banco de semillas e invasión de *Ligustrum lucidum* en bosques costeros de la provincia de Buenos Aires. Argentina. [Composition of the seed bank and invasion of Ligustrum lucidum in coastal forests of the province of Buenos Aires.] *Bosque* (Valdivia) 37(3): 581-590 https://dx.doi.org/10.4067/S0717-92002016000300015.

FAO (2001) State of the World's Forests 2001, *Organización de la Naciones Unidas para la Agricultura y la Alimentación.* http://www.fao.org/forestry/fo/sofo/SOFO2001/sofo2001-e.stm. Accessed: 30 January 2018.

FAO (2016) State of the World's Forests 2016. Forests and agriculture: land-use challenges and opportunity. *Organización de la Naciones Unidas para la Agricultura y la Alimentación* http://www.fao.org/publications/sofo/2016/en/. Accessed: 30 January 2018.

FARN (2010) Soy and conservation of native forest. Fundación Ambiente y Recursos Naturales, http://www.ecosystemalliance.org/sites/default/files/documents/Soy%20Case%201%20-%20Salta%20Argentina.pdf.

FRA (2000) Evaluación de los recursos forestales mundiales 2000. [Evaluation of global forest resources 2000.] *Organización de la Naciones Unidad para la Agricultura y la Alimentación.* http://www.fao.org/forestry/fra/86624/es/.

Frangi P., Sharry S., Abedini W. (1998) *In vitro* Organogenesis of *Schinus molle* var. areira. *Proceedings of 7th Latinoamerican Botany Congress y XIV Mexican Botany Congress.* México. pp 32.

Gabay M., Bessonart S., Barros S. (2011) Latin America −Argentina, Bolivia and Chile. In: Parrota JA, Trosper RL (eds.), *Traditional Forest Related Knowledge Sustaining Communities, Ecosystems and Biocultural Diversity.* Springer. pp. 79-117.

Gattuso M., Rodríguez C., Santecchia S., López E., Martínez S., Zacchino, S. (1998) Estudios *in vitro* de la actividad antifúngica de rizomas y frutos de *Phytolacca tetrámera* Hauman. [In vitro studies of the antifungal activity of rhizomes and fruits of *Phytolacca tetrámera* Hauman.] *Actas VI Simposio Argentino de Farmacobotánica, Argentina,* p. 25.

Goya J., Placi L., Arturi M., Brown A. (1992) Distribución y características estructurales de los talares de la reserva de Biósfera "Parque Costero del Sur". [Distribution and structural characteristics of the talares of the Biosphere Reserve "Parque Costero del Sur".] *Agronomics Faculty Journal. La Plata* 68: 53-64.

Guido A., Rivera S., Rivas M., Ruscitti M., Marinucci L., Galiucci E., Abedini W. (1999) Banco de germoplasma de especies forestales nativas de la Provincia de Buenos Aires. [Germplasm bank of native forest species of the Province of Buenos Aires.] *Actas Jornadas Regionales sobre Estrategias de Conservación de Fauna y Flora Amenazadas. Argentina.*

He J., Yang H., Jamnadass R., Xu J., Yang Y. (2012) Decentralization of tree seedling supply systems for afforestation in the west of Yunnan Province, China. *Small-Scale For.* 11: 147–166.

Hoffmann P., Blum C., Velazco S., Gill D., Borgo A. (2015) Identifying target species and seed sources for the restoration of threatened trees in southern Brazil. *Fauna & Flora Internationals* Oryx 1-6.

Kalesnik F., Valles L., Quintana R., Aceñolaza P. (2008) Parches Relictuales de Selva en Galería (Monte Blanco) en la región del Bajo Delta del Río Paraná. [Relictual Patches of Monte Blanco Forest en Delta del Río Paraná región] *Serie Misc. INSUGEO-CONICET* 17:169 – 193.

Knowles O.H., Parrotta J.A. (1995) Amazonian forest restoration: an innovative system for native species selection based on phenological data and field performance indices. *Commonw for Rev*: 230-243.

Lahitte H.B., Hurrell J.A. (1994) Flora arbórea y arborescente de la Isla Martín García. Nativas y Naturalizadas. Programa Estructura y Dinámica y Ecología del No Equilibrio. [Arboreal and arborescent flora of Martín García Island. Native and Naturalized Program Structure and Dynamics and Ecology of Non-Balance.] *Comisión de Investigaciones Científicas (CIC). Serie Informe* 47:27-29..

Lewis J.P., Noetinger S., Prado D., Barberis I. (2009) Woody vegetation structure and composition of the last relicts of Espinal vegetation in subtropical Argentina. *Biodiversity and Conservation* https://doi.org/10.1007/s10531-009-9665-8.

Lloyd G., McCown B. (1981) Commercially-feasible micropropagation of mountain laurel, *Kalmia latifolia*, by use of shoot-tip culture. *Proc. Intl. Plant Prop. Soc.* 30: 421–427.

Louman B., DeClerck F., Ellati M., Finegan B., Thompson I. (2010) Forest Biodiversity and Ecosystem Services: Drivers of Change, Responses and Challenges Convening. In: *Forests and society – responding to global drivers of change global environmental changes*. International Union of Forest Research Organizations (Ed). IUFRO World Series Vol. 25, Vienna, Austria.

Lu Y., Ranjitkar S., Xu J.C., Ou X.K., Zhou Y.Z., Ye J.F., Wu X.F., Weyerhaeuser H., Jun He J. (2016) Propagation of Native Tree Species to Restore Subtropical Evergreen Broad-Leaved Forests in SW China. *Forests* 7(1): 12. https://doi.org/10.3390/f7010012.

Selection and Propagation of Native Tree Species… 263

Luft J. (2013) *Moving floors: the obstacles to guaranteeing environmental protection of native forests in the context of Argentina's federalism* (Ph.D Thesis) Faculty of the Graduate School of Arts and Sciences of Georgetown University.

Marinucci L., Abedini W., Pariani S., Villarreal B., Bisio C., Sharry S. (2005) Preliminary test for the *in vitro* induction of morphogenesis with *Acacia caven* (Mol.) Mol. *The International Forestry Review* 7 (5):53.

Marinucci L., Ruscitti M., Abedini W. (2003) Morfogénesis *in vitro* en Leguminosas forestales nativas de la República Argentina. [In vitro morphogenesis in native tree legumes from the Argentine Republic.] *Rev Fac Agron La Plata* 105 (2): 27-36.

Merritt D.J., Dixon, K.W (2014) Seed availability for restoration. In: Bozzano M., Jalonen R., Thomas E., Boshier D., Gallo L., Cavers S., Bordács S., Smith P., Loo J.. *Genetic considerations in ecosystem restoration using native tree species.* State of the World's Forest Genetic Resources – Thematic Study. FAO/Bioversity International. pp 281.

Murashige T., Skoog F. A revised medium for rapid growth and bioassays with tobacco tissue cultures. *Physiol. Plant.* 15:473-97, 1962.

Nughes L., Colares M., Hernández M., Arambarri A. (2013) Morfo-anatomía de las hojas de *Celtis ehrenbergiana* (Celtidaceae) desarrolladas bajo condiciones naturales de sol y sombra. *Bonplandia* 22(2):159-170. [Morpho-anatomy of the leaves of Celtis ehrenbergiana (Celtidaceae) developed under natural conditions of sun and shade *Bonplandia* 22(2):159-170].

Oldfield S., Lusty C., MacKinven A. (1998) The World List of Threatened Trees. *UNEP-WCMC*, World Conservation Press. pp. 628.

Painting K., Perry M., Denning R., Ayad W. (1993) *Guía para la documentación de recursos genéticos.* [*Guide for the documentation of genetic resources.*] Consejo Internacional de Recursos Fitogenéticos, Roma. pp 332.

Parodi L. R. (1940) Los bosques naturales de la *Prov. De Bs. As. Anales Acad. Nac. Cienc. Exact. Fís. y Nat.* 7: 97-90. [Native forests of Buenos Aires province. Proceedings National science Academy 7: 97-90].

Pritchard H., Moat J., Ferraz J., Marks T., Camargo J., Nadarajan J., Ferraz I. (2014) Innovative approaches to the preservation of forest trees. *Forest Ecology and Management* 333: 88-98. https://doi.org/10.1016/j.foreco.2014.08.012.

Ramilo D.I., Abedini W.I. (2007) Propagación vegetativa de *Celtis tala* Gill. Ex Planch por enraizamiento de estacas de madera semilignificada. *XXI Jornadas Forestales de Entre Ríos*. Argentina. [Vegetative propagation of Celtis tala Gill. Ex Planch by rooting wooden stakes semilignified. XXI Jornadas Forestales de Entre Ríos. Argentina].

Ramilo D.I., Abedini W.I (unpublished). *Propagación vegetativa de Scutia buxifolia Reissek por enraizamiento de estacas de madera semilignificada. Informe interno.* CEPROVE. Facultad de Ciencias Agrarias y Forestales – UNLP. [*Vegetative propagation of Scutia buxifolia Reissek by rooting wooden stakes semilignified, internal report.* CEPROVE. School of Agriculture and Forestry Sciences].

Ribichich A., Protomastro J. (1998) Woody vegetation structure of xeric forest stands under different edaphic site conditions and disturbance histories in the Biosphere Reserve Parque Costero del Sur, Argentina *Plant Ecology* 139 (2):189–201.

Rivas M.C., Abedini W. (1996) Rapid clonal propagation of *Phyllantus sellowianus* (Sarandí blanco). *Third International Symposium on in vitro culture and Horticultural Breeding*. Jerusalem, Israel.

Rivas M.C., Abedini W.I., Sharry S.E. (2004) Forest Genetic Resources in the Buenos Aires Province, Argentina: characterization, conservation and propagation. *FAO Recursos Genéticos Forestales* 31:57-60.

Romero J. (2012) *Forest conservation in Argentina: early analysis of the Forest Law implementation in the Chaco Ecoregion*. Thesis. University of British Columbia.

Roussy L., Pinciroli L., Briones V., Sharry S., Ciocchini G., Abedini W. (2011) Enraizamiento de estaquillas de diferente diámetro de *Citharexylum montevidense* (Spreng) Moldenke, especie forestal nativa de la flora bonaerense, Argentina. Perú. LIMA. *V Congreso Forestal Latinoamericano. CONFLAT.* Universidad de La Molina. [Rooting of cuttings of different diameter of Citharexylum montevidense (Spreng)

Selection and Propagation of Native Tree Species... 265

Moldenke, forest species native to the flora of Buenos Aires. V Latinamerican Forestry Congress, Lima, Peru. CONFLTA. La Molina University].

Ruscitti M, Abedini W (1996) Clonación *in vitro* de ecotipos de *Erythrina crista-galli* resistentes a contaminación ambiental. *XVI Congreso de Fitogenética. Sociedad Mexicana de Fitogenética (SOMEFI). Instituto de Recursos Genéticos y Productividad del Colegio de Postgraduados Montecillo, México.* [In vitro cloning of Erythrina crista-galli ecotypes resistant to environmental contamination. XVI Congress of Phytogenetics. Mexican Society of Phytogenetics (SOMEFI). Institute of Genetic Resources and Productivity of the Postgraduated School of Montecillo, Mexico].

Sailer D. 2006. Propagating native plants. In: I Ho´ōla I Ka Nahele: *To Heal A Forest A Mesic Forest Restoration Guide for Hawaii.* pp 1-39.

Secretaría de Ambiente y Desarrollo Sustentable Argentina (2014) *Estimaciones Unidad del Sistema de Evaluación Forestal, Dirección de Bosques.* http://estadisticas.ambiente.gob.ar/archivos/web/Indicadores/ file/multisitio/fichas/082015/15-%20Superficie%20de%20Bosque% 20Nativo_2015.pdf. Accessed: 1 February 2018. [*Estimates Forest Appraisal System Unit, Forest Directorate.* http://estadisticas. ambiente.gob.ar/archivos/web/Indicadores/file/multisitio/fichas/08201 5/15-%20Superficie%20de%20Bosque%20Nativo_2015.pdf. Accessed: 1 February 2018.].

Sharry S., Abedini W., Basiglio Cordal M., Briones V., Roussy L., Stevani R., Galarco S., Adema M. (2011) Food and medicinal value of some forest species from Buenos Aires (Argentina) *Emirates J Food Agric* 23(3): 222-236.

Sharry S., Adema M., Basiglio M., Villarreal B., Nikoloff N., Briones V., Abedini W. (2011) Propagation and Conservation of Native Forest Genetic Resources of Medicinal Use by Means of *in vitro* and *ex vitro* Techniques. *Nat. Prod.* 6 (0): 1-4.

Sharry S., Abedini W. (2002) Obtención de callos in *vitro* de *Celtis tala* Gill. Ex Planch. *Actas XI Congreso Italo-latinoamericano de Etnomedicina.* Italia. [In vitro plants and callus obtaining of *Celtis tala* Gill. Ex Planch. Proceedings XI Italain-latinoamerican Etnomedicine Congress. Italia].

Sharry S., Abedini, W. (2014) Estrategias biotecnológicas aplicadas en la conservación de especies forestales nativas bonaerenses. *Agusvinnus* 0: 1- 30. [Biotechnological strategies applied in the conservation of native forest species in Buenos Aires Agusvinnus 0: 1- 30].

Sharry S., Lede S., Abedini W (1997) *In vitro* tissue culture of de *Schinus molle* var. areira (aguaribay). *Proceedings II World Congress of Aromatic and Medicinal Plants for the Welfare of Humanity.* Argentina. p 166.

Stupino S., Arturi M., Frangi J. (2004) Estructura del paisaje y conservación de los bosques de *Celtis tala* Gill ex Planch del NE de la provincia de Buenos Aires. *Rev Fac Agron La Plata* 105:37-45. [Landscape structure and forest conservation of *Celtis tala* Gill ex Planch in NE of Buenos Aire province. *Rev Fac Agron La Plata* 105:37-45.]

Tinto J. (1977) Utilización de los Recursos Forestales Argentinos. Instituto Forestal Nacional. Subsecretaría de Recursos Naturales Renovables y Ecología. Ministerio de Economía. *Folleto técnico forestal* 41:68. [Utilization of the Argentine Forest Resources. National Forest Institute. Undersecretariat of Renewable Natural Resources and Ecology. Ministry of Economy. *Forest technical brochure* 41:68].

Toribio M, Celestino C (2000) El uso de la biotecnología en la conservación de recursos genéticos forestales. In: Gil LA, Alía R (eds). Conservación de Recursos Genéticos Forestales. Investigación Agraria. Sistemas y Recursos Forestales. *Fuera de Serie* 2: 249-260. España. [The use of biotechnology in the conservation of forest genetic resources. In: Gil LA, Alía R (eds.) Conservation of Forest Genetic Resources. Agricultural Research. Forest Systems and Resources. *Fuera de Serie* 2: 249-260].

Torretta J., Basilio A. (2013) Pollen dispersion and reproductive success of four tree species of a xerophytic forest from Argentina. *Rev de Biología.*

Zacchino, S. 2000. Compuestos Antifúngicos naturales con actividad antifúngica promisoria. *J. Ethnopharm.* 1:29-34. [Natural antifungal compounds with promising antifungal activity *J. Ethnopharm.* 1:29-34].

In: Forest Conservation
Editor: Pedro V. Eisenlohr

ISBN: 978-1-53614-559-5
© 2019 Nova Science Publishers, Inc.

Chapter 10

CONCEPTS AND METHODS IN ENVIRONMENTAL SUITABILITY MODELING, AN IMPORTANT TOOL FOR FOREST CONSERVATION

João Carlos Pires-Oliveira[1], Leandro José-Silva[1], Diogo Souza Bezerra da Rocha[2] and Pedro V. Eisenlohr[1,*]

[1]Laboratory of Ecology, State University of Mato Grosso, Alta Floresta, MT, Brazil
[2]National School of Tropical Botany, Botanic Garden of Rio de Janeiro, Rio de Janeiro, RJ, Brazil

INTRODUCTION

The biological diversity of our planet is so high that we do not have yet a precise number to represent it (Wilson 2000; Wilson 2003), even with the

---

* Corresponding Author Email: pedro.eisenlohr@unemat.br.

advances in technologies to make estimates in this field (Guralnick et al. 2007). There are organisms living in the most unlikely places, such as the polar ice caps, the extremely arid deserts (i.e., the Atacama Desert) and the deepest oceans (Wynn-Wilhams 1996; Willerslev et al. 1999; Warren-Rhodes et al. 2006), but it is in the tropical region that most of this biodiversity is concentrated (Jenkins 2003). In fact, biodiversity is present in all its greatness in tropical forests (e.g., Slik et al. 2015), such as the Amazon Rainforest in the American continent, the Rainforest of Congo and the Rainforest of Southeast Asia.

Great efforts have been invested to fill knowledge gaps regarding the geographical distribution of species in tropical forests (Kerr 1997; Myers et al. 2000; Siqueira et al. 2009; Oliveira-Filho 2017). In the context of the conservation of tropical biodiversity, knowledge about the ecological and distribution patterns of species is of particular importance, since the risk of neglecting a key species due to lack of data can significantly compromise decision-making. In this context, a large number of tools have been developed, such as the Environmental Suitability Modeling (ESM) techniques, in order to predict or project the potential distribution of different species (Elith et al. 2006; Liu et al. 2011; Zhang et al. 2015). These tools use predictive variables (usually abiotic data) that are relevant to determine the species occurrence, which are combined with georeferenced data to generate an environmental suitability model. It is possible to project this model back into the geographic space, thus indicating suitability areas for the occurrence of the species based on the range of environmental conditions in which it occurs (Elith et al. 2006; Elith & Leathwick 2009; Peterson et al. 2011). In short words, ESM 'transforms' geographic distribution of occurrence points into a distribution surface, or suitability area, based on environmental variables. Even the models presenting a considerable degree of uncertainty, they still represent an excellent alternative or a complement to the *in situ* studies in order to describe the geographical ranges of the species (Smith et al. 2006; Liu et al. 2011; Gastón et al. 2014; Zhang et al. 2015).

The geographic distribution of a given species is related to its niche (i.e., requirements necessary to establish persistent populations), but also to factors related to the capacity and possibility of dispersion, as well as interspecific relations (i.e., competition, predation, parasitism) and historical factors (Brown et al. 1996; Chave et al. 2002). The ESMs seek to represent this universe of conditions and resources by means of a subset of environmental layers, but they do not reveal much about the movement of individuals, since dispersal data are rare or non-existent for most species. The ESM finds support in Hutchinson's niche theory (1957), which is its central pillar. This theory is based on the current ecological factors in which the species survive, grow and reproduce, or, in other words, the conditions that the species tolerates and the resources that supply its requirements. Based on this prior knowledge, space projections are generated to identify suitable areas for the occurrence of the species (Soberon 2007; Pearman et al. 2008). Thus, two fundamental assumptions must be considered when using ESMs: 1) the species or ecological group considered in the study is in equilibrium with the environment (de Marco-Junior et al. 2008), and 2) the niche of these groups is conserved in space and time, since environmental conditions are considered the main factor limiting the distributions of species in general (Pearson and Dawson 2003; Soberón 2007; Soberón and Nakamura 2009).

The ESM has been extensively used with multiple applications, including the field of conservation science, in studies that aim to better understand the patterns of species distribution on the Earth's surface, considering scenarios of the present, past and future (Heikkinen et al., 2006; Franklin et al., 2013; Sillero et al., 2013). This is a powerful tool, for example, as a starting point for the selection of areas for exploratory excursions and inventories of biodiversity (Jimenez-Valverde et al., 2008). In addition, one can make predictions of changes in patterns of species distribution in response to the effects of global climate change (Pacifici et al., 2015). However, all the advantages offered by ESMs lose their value if some care is not taken. Such care should be taken under both conceptual and methodological points of view. Despite the extensive literature that discusses the application or even the designation of these models (Franklin 2009; Peterson and Soberón 2012; Guisan et al. 2017), researchers wishing to enter the universe of environmental suitability modeling may miss a text that addresses the central conceptual and methodological issues or even a more concise text, which should expected to be not exhaustively dedicated to formulas.

Limitations on conceptual and methodological issues that support the use of ESMs are still poignant. Once this knowledge gap is identified, we intend to discuss central issues of the use of this tool, highlighting what we will call as "pitfalls of environmental suitability modeling". Our focus is to present a theoretical framework for researchers wishing to have preliminary lessons on ecological modeling. Considering this, we will first separate the conceptual issues from the methodological ones to facilitate our approach. We will also present the most frequent errors found in papers published in scientific periodicals, and some ways to avoid such errors or to minimize their effects.

CONCEPTUAL ISSUES

Concepts are fundamental parts of science, which must be as clear and universal as possible, so that they can be used in all places and fields of science without loss or distortion in their meaning. Therefore, before we consider methodological issues, it is necessary to clarify some conceptual problems that frequently appear in papers that use ESMs. These errors can lead to misconceptions and erroneous interpretations of the results, thereby rendering the final product unreliable.

The first conceptual issue is related to the name of the modeling tool. Before choosing a title or even writing a text about ESMs, it is necessary to think on what is being modeled. Without a clear definition of this, the readership will be confused about what those results represent and what are their practical applicability. This doubt may lead to misuse of the results presented by the models, simply because there is no consensus on how they should be called (Guisan et al. 2017). In order to standardize terms on suitability models, some scientists have attempted to propose a term covering all types of work that use the ESMs (Peterson and Soberón 2012; Guisan et al. 2017). This fact results in varied ways of handling the process

Concepts and Methods in Environmental Suitability Modeling ... 273

of ecological modeling (Austin 2002; Soberón and Peterson 2005). In a quick search on the popular scientific work indexer 'Google Scholar' (https://scholar.google.com/) it is common to find works that use the terms "Species Distribution Modeling/Species Distribution Modelling", "Ecological Niche Modeling/Ecological Niche Modelling", "Environmental Niche Modeling/Environmental Niche Modelling" and more recently, "Habitat Suitability Modeling/Habitat Suitability Modelling".

Species Distribution Modeling/Species Distribution Modelling - We obtained 211 results in a quick search using the keywords "Species Distribution Modeling/Modeling" as title words in Google Scholar, where we filtered the papers within the range between 2010 and 2017. We chose this period because the great rise of the ESMs has started on 2010, which can be confirmed by the growing number of published works in the field, and 2017 was chosen as upper limit year in order to avoid interference from papers published in the current year (2018). Here we define species distribution modeling as the real probability of finding an individual of the modeled species, considering the conditions and resources necessary for its persistence in the area and also its capacity of movement. This includes the definition used by Franklin et al. (2009) and Guisan et al. (2017).

Ecological Niche Modeling/Ecological Niche Modelling - We obtained 209 records by using the above keywords during our search. In fact, the term "Ecological Niche Modeling/Ecological Niche Modelling" was popularized after the study of Peterson and Soberón (2012), which presented an excellent discussion on which terminology is the most appropriate to refer to suitability models; however, even before 2012, this denomination had already being used (Peterson 2003; Barve et al. 2011). When we use the word 'niche' in the current text, we follow Hutchinson (1957), i.e., we consider niche as a hypervolume composed of n dimensions where the species meets the set of conditions and resources necessary for its survival. Thus, when we refer to ESMs as ecological niche models, we would be intrinsically assuming that the results of these models are representative of the conditions and resources that the species needs, and this is not necessarily true.

Environmental Niche Modeling/Environmental Niche Modelling - This was the least used term to describe the ESMs in the article titles, appearing only nine times in our research. This low representativeness is somewhat incomprehensible, since among all the terms mentioned this is the closest to a coherent denomination for the ESMs. Such definition contains a key term to describe the ESMs: the environment. However, we emphasize that the use of the word 'niche' in this context can lead to inaccuracies of interpretation, since niche is something very complex to be expressed only in environmental terms, and the caveats made in the previous item on employment of the word 'niche' also apply here.

Habitat Suitability Modeling/Habitat Suitability Modelling - The searches for terms referring to habitat suitability returned 88 results. This is the latest proposed term to describe the ESMs and became more popular after the publication of Guisan et al. (2017). These authors developed an excellent discussion about the correct designation for suitability models, but they could be questioned in defining the models as species habitat projections. According to Oxford Dictionary (OD), the definition of habitat is the representation of the "natural environment", natural territory, habitation or dwelling of an organism (i.e., animal or plant). So, if we consider the OD definition, the use of 'habitat' would not be a suitable choice to describe what is in fact produced by the ESMs. The models infer nothing about the natural environment or housing of organisms. What they actually produce are estimates of the appropriate environmental conditions based on the occurrence records.

Among the 517 papers, 40.81% used the term "species distribution modeling" in their titles; 40.43% worked with "ecological niche modeling"; 17.02% opted for "habitat suitability modeling", and only 1.74% used "niche modeling" terminology. We consider the latter a more suitable approach than the previous ones, but the choice of the word "niche" should be used with extreme caution and with explanation of what is being considered "environmental niche". The concept of niche (Hutchinson 1957; Chase and Leibold 2003) is a central factor for modeling (Franklin et al. 2009), but what the models present are, in fact, areas with environmental suitability, i.e., areas presenting environmental conditions (including climatic

Concepts and Methods in Environmental Suitability Modeling ... 275

conditions) for the establishment/ development of a given species, similar to those where known occurrence records are present.

BAM Diagram

The values described above do not describe the actual uses of each of the terms, but we can see which are the most used. Each of these terms should be used under a very narrow range of conditions (Peterson & Soberón 2012; Guisan et al. 2017). In addition, most of these studies use mainly precipitation and temperature data as predictors of the models (Araújo and Luoto 2007; Vogler et al. 2013; Váz and Nabout 2016), and these descriptors do not represent the complexity of the environmental niche, the ecological niche or the species distribution (Jiménes-Valverde et al. 2008). Thus, the models only indicate areas with environmental conditions that are similar to those where the species is located, considering only the variables inserted in the model to indicate areas where the species could be established. Any other kind of interpretation about what the models represent is not necessarily correct. The BAM diagram, initially proposed by Soberón and Peterson (2005), is a very simple and suitable way to represent the main factors that are present when working with species distribution (Figure 1). We will make use of this diagram here, because we have a very useful tool to represent factors that are fundamental for the ecological knowledge about the species and, consequently, for the production of ESMs.

Among the factors presented in the diagram of Figure 1, only the intersection between two of them (G) is used in the construction of the ESMs. As previously mentioned, the models only contain environmental variables (abiotic - A) and, in some rare cases, using dispersion data (M - movement) is also possible (Soberón & Peterson 2005).

Another very frequent conceptual error appears when the authors treat the suitability values generated by the models as probability of presence of the species, that is, as a probability value of finding an individual of the species modeled. This interpretation is wrong. The models present projections derived from an environmental space, i.e., they do not represent the conditions in the geographic space. In the next step, the data associated with those points are extracted from the provided variables (environmental layers) and projections are made based on the extracted data (Figure 2).

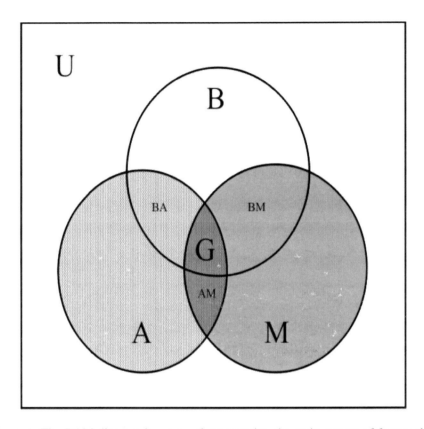

Figure 1. The BAM diagram is a way of representing the main groups of factors that influence species distribution patterns, where *B* represents those provided by *biotic* interactions, *A* represents the *abiotic* conditions exerted by the environment and *M* consists of the influence exerted by the displacement capacity of the species, or *movement*. All these components are within a space called "universe", represented here by the letter U. The BA intercept represents favorable biotic and abiotic conditions, but areas are inaccessible (-M); BM consists of favorable and accessible biotic conditions, but favorable abiotic conditions are not present; AM is the description of sites with favorable and accessible abiotic conditions, but favorable interactions with other species are not present. Finally, the area represented by G consists of the realized niche, i.e., the area that is actually occupied by the individual (Soberón & Peterson 2005).

Concepts and Methods in Environmental Suitability Modeling ...   277

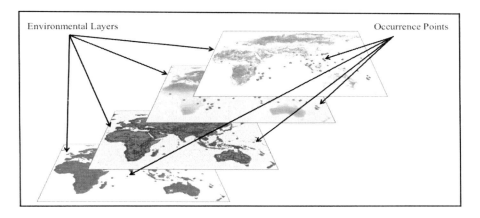

Figure 2. Environmental layers with occurrence points. Each point brings the data of the corresponding environmental layers; these data will be used by the algorithms to build ESMs.

Then, these data are projected into the environmental space, where the data corresponding to each of the variables will be treated in a two-dimensional plane. Then, the algorithms correlate the points with the environmental data associated with them to find similar environmental conditions (Figure 3).

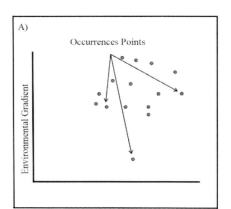 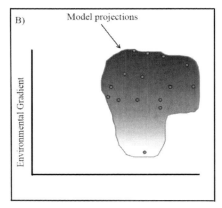

Figure 3. Diagram representing the environmental space along an environmental gradient (Temperature in the X axis and Precipitation in the Y axis) and the projections of the model referring to these points. A) Occurrence points arranged in a two-dimensional plane; B) Occurrence points arranged in (and influenced by) an environmental gradient.

Methodological Issues

As mentioned earlier, the use of ESMs has increased, but even with these advances, one should not neglect central aspects of the methodology. The first aspect that we will address here is related to the compilation of species occurrence data. The simple act of searching occurrence data on the web or even in natural history museums and other individual collections does not guarantee a robust data set. Eisenlohr and Oliveira-Filho (2015) present some techniques to increase the reliability of metadata used in the construction of the models. We will mention here some of these techniques, propose others and make some suggestions to avoid problems related to occurrence data and to the predictor variables of the models.

Occurrence Records

One of the first steps for those who want to use ESMs to answer their questions is to find occurrence data from species or even interest groups. However, this may become a more arduous task than originally planned, since in works that intend to investigate rare species there are not always occurrence records of these species due to the difficulty of finding specimens in databases.

Data on the occurrence of species are frequently searched in large databases of species such as GBIF (https://www.gbif.org/). There are also relevant national databases that should be consulted, such as SpeciesLink (http://www.splink.org.br) for Brazilian biota. Using all the occurrence records obtained from such databases to construct ESMs is a serious problem (Beck et al. 2014). Another problem is that we cannot always be sure that the data recorded in the database correspond to the same species we are studying, i.e., we can have a taxonomic bias (Troudet et al. 2017). An alternative to circumvent this issue is to search for occurrence data in databases that provide images of the exsiccates and/or the determiner. If these data are insufficient or if a more complete sampling is required, searches may be carried out in specialized journals sheltering the group that will be modeled (species, genus or even an ecological group). While the database gains reliability, there is no guarantee that the researcher will have a clean and fully reliable taxonomic data, but this procedure reduces the likelihood of working with heavily biased data.

Even when we have confidence in the identification of the species and/ or we can verify if this identification is correct, we still have the possibility that the georeferencing of the occurrence points, when present, is not very precise (Beck et al. 2014). In this situation we can adopt some procedures to gain reliability that the georeferenced data corresponds to the informed locality. One of these procedures is to verify if the collectors have made a description of the location where the sampling was obtained online, since, even with poor-quality georeferencing, we can still retrieve the coordinates of the locality through softwares or platforms such as Google Earth (Guralnick et al. 2007). This is not the best way to perform a data collection or cleaning, but in case of a few points it may be the only way to make the work viable. Another way to have more confidence in our data is to follow the recommendation of Eisenlohr and Oliveira-Filho (2015), which consists of plotting occurrence points in a shapefile using a GIS environment (i.e., QGIS, DIVA-GIS or even functions in R environment). We also indicate the use of environmental filters, which can reduce sampling bias (Varela et al. 2014).

Explanatory Variables

After completing the process of cleaning the occurrence data, it is time to choose which variables will be used as predictors of the ESMs. This step may seem trivial compared to the ones described above, but this is not true. These models must present relevant information on the response curves of the species, which are presented here in the form of environmental suitability of the species (Austin 2002; Chunco et al. 2013). For this reason, the use of variables that have a relevant influence on the processes that condition the distribution is substantial. In this section we will present some caution that we should take when selecting variables for model construction, and will also present methods to deal with each of the possible problems.

The choice of how many variables should be used in our models plays a central role in building the ESMs, and there is no "magic formula" to perform such a task. Researchers should use a set of variables with the goal of representing the fundamental niche of the species as fully as possible (Peterson et al. 2011). However, using a very large set of environmental variables can result in overfitting and, thus, models with high complexity and difficult interpretation (Guisan and Zimmermann 2000).

Another common concern is related to collinearity or multicollinearity, that is, a high degree of redundancy among predictor variables (Dormann et al. 2013). Model generated with very collinear variables shows inflation of commission errors, i.e., the model predicts an area as suitable for the species when it is not (Pearson 2007). For these reasons, we recommend never using variables without verifying the collinearity between the variables. This verification can be performed through VIF (*variance inflation factor*) calculation (Booth et al. 1994). VIFs above 20 indicate strong collinearity, but VIFs above 10 should be examined and avoided (Borcard et al. 2011). However, if it is necessary or important to insert many variables, we suggest using the PCA (Principal Components Analysis) axes as predictors, since they are orthogonal and do not present collinearity (Dormann et al. 2013).

It is also preferable that the predictor variables are on the same scale, thus ensuring that the results are actually representing the unit we wish to present (Guisan et al. 2017). Variables at different scales may have their values interpreted differently, because variables with higher numerical values will be treated as variables that exert a greater influence on response variables (Quinn and Keough 2002).

A PROPOSED ESM ROUTINE

In order to achieve a robust, reliable and practical applicability model, we suggest two routines available at <https://github.com/pedroeisenlohr>, which present a sequential and clear developmental way of applying ESMs.

We will present each of the modeling routines separately for better understanding. All routines were fully developed in the R statistical environment (R Development Core Team 2018).

The routines have two versions: one for the development of ESMs for the current time, which has a set of variables compiled from several sources, and the other one developed to create projections for the future (2050 and 2070) based on projections of the fifth report of the Intergovernmental Panel on Climate Change - IPCC.

In order to construct this framework we used the 'biomod2' package (Thuiller et al. 2013), which currently has 11 algorithms (General Linear Model - GLM; Generalized Additive Model - GAM; Generalized Boosting Model - GBM; Classification Tree Analysis - CTA; Artificial Neural Network - ANN; Surface Envelope – SER (also known as BIOCLIM); Flexible Discriminant Analysis - FDA; Multiple Adaptive Regression Splines - MARS; Random Forest - RF; and Maximum Entropy – MAXENT-Phillips and MAXENT-MAXENT.Tsuruoka (Phillips et al. 2006; Thuiller et al. 2009). The 'biomod2' also provides seven metrics to evaluate the predictive ability of the models: Area Under the Relative Operating Characteristic Curve - AUC, True Skill Statistic - TSS, Success Ratio - SR, False alarm ratio - FAR, Probability of Detection - POD, ACCURACY and Bias Score - BIAS (Thuiller et al. 2016). Only the first two will be used in the proposed routines, since they are the most suitable for the type of modeling suggested here (Allouche et al. 2006; Peterson et al. 2008; van Zonneveld et al. 2014). However, these evaluative metrics should be restricted to cases where many occurrence points (>25) are available, as suggested by Pearson et al. (2007). The predictive ability of models produced from small sets (<25 occurrence points) should be evaluated using a Jackknife (leave-one-out) framework (Pearson et al. 2007).

ESMs under Current Climate

After having verified and corrected possible problems with the occurrence matrix in the modeling routine and selected only spatially unique occurrence records, a Principal Component Analysis (PCA) can be applied. Here, a number of axes representing a very expressive fraction (here, 90%) of the variation contained in those data is retained; these axes will then be used as the predictor variables of the models. Thus, as previously mentioned, we eliminate any collinearity between the variables, since PCA axes are mathematically independent of each other (Dormann et al. 2013). In this routine, a set of approximately 70 variables is used. These variables include climatic, soil, vegetation cover and relief and topography descriptors, but the researcher who intends to use this routine can use the set of variables that best meet his/her work issues.

After all the checks on the occurrence data and the predictor variables, it is time to format the data to start the modeling. In this step we will prepare the data to be used in the models, i.e., we will establish the necessary parameters to perform the modeling. Firstly, we select part of the occurrence data for the construction of the models and part for the purposes of evaluating the models (here we use 70% and 30%, respectively; Fielding et al., 1997). Then we split the set of 10 algorithms into two groups, since three algorithms (GBM, CTA and RF) require different configurations (Barbet-Massin et al. 2012). These authors recommend a calibration for these three algorithms such that the number of pseudo-absences (PAs) must be equal to the number of occurrence records of the species. However, other algorithms (GLM, GAM, ANN, SRE, FDA, MARS and MAXENT) can use a calibration with 10.000 PAs (Barbet-Massin et al. 2012).

At this stage of the routine, models will be built based on the occurrence points and environmental variables provided. After this step, the models are evaluated by two criteria: Area Under the ROC (Relative Operating Characteristic) Curve - AUC and True Skill Statistic - TSS.

After the evaluation we discarded the models that do not present a good performance according to the evaluation metrics, which we define here as those presenting values of TSS <0.4 and AUC <0.8, because they present low accuracy (Allouche et al. 2006; Zhang et al. 2015). The next step is to create the projections, i.e., the maps of environmental suitability. These projections are initially separated by algorithms and "stacked" to create a continuous model of consensus projections.

Concepts and Methods in Environmental Suitability Modeling ... 283

In addition to a continuous model, this routine also provides the possibility of creation of binary models (presence-absence) for those who need to make a "occurs or does not occur" decision. These models are produced with the application of a threshold that determines which areas will be considered as suitable (i.e., values above this threshold), and areas that will be considered unsuitable (i.e., values below the threshold). We opted to use the threshold that maximizes sensitivity and specificity (ROC threshold), which implies a lower overprediction (Liu et al. 2005).

ESMs under Future Climate Scenarios

The second routine of ESMs construction that we will present is capable of generating projections for the years 2050 (2041-2060) and 2070 (2061-2080) using the variables available from the CHELSA 1.2 bioclimatic database (see http://chelsa-climate.org/). These data are constituted from a set of 19 variables, of which 11 are temperature variables and eight are precipitation variables. In this database there are sets of variables that represent the expected climatic conditions for the Earth's future under various scenarios of the Representative Concentration Pathways (RCP) as presented by the fifth IPCC report. Here we use data from seven general circulation models (GCM). GCMs simulate the behavior of planetary atmospheric circulation, and thus can be used to infer large-scale climate responses (Mechoso and Arakawa 2015). The CHELSA database currently contains 30 GCMs, but in this routine we use only 7 (CCMSM4, CMCC-CM, CSIRO-Mk3-5-0, GFDL-CM3, HadGE M2-CC, MIROC5 and MIROC-ESM), since they present the greater variation in the data according to the recommendation available on the CHELSA web page.

In the case of this routine, we only work with the scenario regarding carbon emissions considered pessimistic (rcp 8.5), which becomes closer to the actual observations. This routine follows all steps of checking occurrence data described earlier. The collinearity verification method of environmental variables used here is the Variance Inflation Factor - VIF, but the use of PCA is also possible (e.g., Zwiener et al. 2017). The analysis aims to detect the collinearity between the 19 predictors and to select a subset whose variables present a VIF of less than 10 (Quinn and Keough 2002; Borcard et al. 2011). Double-check should be made on the set of variables to detect variables with redundant information. The first verification of collinearity is performed on the variables associated with the points, and the second, on the extracted variables of 10.000 points randomly distributed throughout the background. Adopting this framework, we control the problem of collinearity between the predictor variables. For the development of these routines, the data available in the CHELSA database (Karger et al. 2017) were used, but with few changes one can use any set of variables that provide projections for the required time interval. The procedure for formatting the data and creating the models are described in section 4.1, and they do not differ at all from the calibration used here. The next step is to evaluate these models; from now on, only models with TSS> 0.4 will be used (Zhang et al. 2015).

Next, the consensual models are generated for each algorithm individually. Then, the projections are generated for a given time (i.e., 2050 and 2070 years). The procedures for generating projections for 2050 and 2070 years do not differ. Finally, these future projections are grouped by algorithms or by GCM, to generate the average consensus of each algorithm or each GCM, binary or continuous.

FINAL CONSIDERATIONS

ESMs are very powerful tools and have considerable flexibility. However, we must be attentive to the concepts employed during the initial phase of ESM production. If ESMs are not built considering a solid theoretical basis, their results will be nothing more than a set of beautiful figures without any relevance to the scientific world. We also believe that Environmental Suitability Modeling is the best nomenclature to describe what such models are in fact.

We also highlight that ESMs are developed using subsets of predictor variables and correlate species occurrence data with environmental data to predict areas with environmental suitability for this/these species. In this way, output maps are directly related to the selection of these variables. However, even with reliable presence data, selecting variables is not a simple task, since each species needs a plethora of both ecological conditions and resources, and this information is not always available. Thus, caution must be taken to minimize the effects of these factors.

Here we listed errors that, in our opinion, are the most common found in the production of ESMs. However, many others may occur. Thus, it is of utmost importance that all those working with biological conservation or any other related area that wish to venture into the field of the ESMs have their guide questions well established. Without this, the chance of extracting useless solutions from this immensity of forms and formulas is too great, making the researcher more prone to fall into pitfalls such as the ones mentioned here.

To conclude, we presented a robust routine that acts against the possible methodological errors addressed here. We emphasize that the ESMs are very welcome tools to help in the biodiversity conservation context, not only for tropical forests, but also for any other biome.

REFERENCES

Allouche, O., Tsoar A., Kadmon R. (2006). Assessing the accuracy of species distribution models: Prevalence, kappa and the true skill statistic (TSS). *J Appl Ecol* 43:1223–1232.

Araújo, M. B., Luoto, M. (2007). The importance of biotic interactions for modelling species distributions under climate change. *Glob Ecol Biogeogr* 16:743–753.

Austin, M. P. (2002). Spatial prediction of species distribution: An interface between ecological theory and statistical modelling. *Ecol Modell* 157:101–118.

Barbet-Massin, M., Jiguet, F., Albert, C. H., Thuiller, W. (2012). Selecting pseudo-absences for species distribution models: How, where and how many? *Methods Ecol Evol* 3:327–338.

Barve, N., Barve, V., Jiménez-Valverde, A., et al. (2011). The crucial role of the accessible area in ecological niche modeling and species distribution modeling. *Ecol Modell* 222:1810–1819.

Beck, J., Böller, M., Erhardt, A., Schwanghart, W. (2014). Spatial bias in the GBIF database and its effect on modeling species' geographic distributions. *Ecol Inform* 19:10–15.

Booth, G. D., Niccolucci, M. J., Schuster, E. G. (1994). Identifying proxy sets in multiple linear-regression - an aid to better coefficient interpretation. *USDA For Serv Intermt Res Stn Res Pap* 1–13.

Borcard, D., Gillet, F., Legendre, P. (2011). *Numerical Ecology with R, 1ª*. Springer US, New York.

Brown, J. H., Stevens, G. C., Kaufman, D. M. (1996). The geographic range: size, shape, and internal structure. *Annu Rev Ecol Syst* 27:597–623.

Chase, J. M., Leibold, M. A. (2008). Ecological niches. *Linking classical and contemporary approaches.*

Chave, J., Leigh, E. G. (2002). A Spatially Explicit Neutral Model of β-Diversity in Tropical Forests. *Theor Popul Biol* 62:153–168.

Chunco, A. J., Phimmachak, S., Sivongxay, N., Stuart, B. L. (2013). Predicting environmental suitability for a rare and threatened species (Lao Newt, *Laotriton laoensi*) using validated species distribution models. *PLoS One* 8:1–13.

de Marco-Júnior, P., Diniz-Filho, J. A. F., Bini, L. M. (2008). Spatial analysis improves species distribution modelling during range expansion. *Biol Lett* 4:577–580.

Dormann, C. F., Elith, J., Bacher, S., et al. (2013). Collinearity: A review of methods to deal with it and a simulation study evaluating their performance. *Ecography (Cop)* 36:027–046.

Eisenlohr, P. V., Oliveira-Filho, A. T. (2015). Obtaining metadata for phytogeographic of synthesis works and the NeotropTree database as a case study. In: Eisenlohr PV, Felfili JM, Melo MM de RF de, et al. (eds) *Phytosociology in Brazil: Methods and Case Studies*, Vol. 2. Ed. UFV, Viçosa, pp 387–411.

Elith, J., Graham, C., Anderson, R., et al. (2006). Novel methods improve prediction of species' distributions from occurrence data. *Ecography (Cop)* 29:129–151.

Elith, J., Leathwick, J. R. (2009). Species Distribution Models: Ecological explanation and prediction across space and time. *Annu Rev Ecol Evol Syst* 40:677–697.

Fielding, A. H., Bell, J. F. (1997). A review of methods for the assessment of prediction errors in conservation presence/absence models. *Environ Conserv* 24:38–49.

Franklin, J. (2009). *Mapping species distributions,* 1ª. Cambridge University Press, Cambridge.

Franklin, J., Davis, F. W., Ikegami, M., et al. (2013). Modeling plant species distributions under future climates: How fine scale do climate projections need to be? *Glob Chang Biol* 19:473–483.

Gastón, A., Garcia-Vinas, J. I., Bravo-Fernandez, A. J., et al. (2014). Species distribution models applied to plant species selection in forest restoration: are model predictions comparable to expert opinion? *New For* 45:641–653.

Guisan, A., Thuiller, W., Zimmermann, N. E. (2017). *Habitat Suitability and Distribution Models.* Cambridge University Press, Cambridge.

Guisan, A., Zimmermann, N. E. (2000). Predictive habitat distribution models in ecology. *Ecol Modell* 135:147–186.

Guralnick, R. P., Hill, A. W., Lane, M. (2007). Towards a collaborative, global infrastructure for biodiversity assessment. *Ecol Lett* 10:663–672.

Heikkinen, R. K., Luoto, M., Araújo, M. B., et al. (2006). Methods and uncertainities in bioclimatic envelope modelling under climate change. *Prog Phys Geogr* 30 (6):751–777.

Hutchinson, G. E. (1957). Population studies: animal ecology and demography. *Bull Math Biol* 53:415–247.

Jenkins, M. (2003). Prospects for Biodiversity. *Science* (80-) 302:1175–1177.

Jiménez-Valverde, A., Lobo, J. M., Hortal, J. (2008). Not as good as they seem: The importance of concepts in species distribution modelling. *Divers Distrib* 14:885–890.

Karger, D. N., Conrad, O., Böhner, J., et al. (2017). Climatologies at high resolution for the earth's land surface areas. *Sci Data* 4:1–20.

Kerr, J. T. (1997). Species richness, endemism, and the choice of areas for conservation. *Conserv Biol* 11:1094–1100.

Liu, C., Berry, P. M., Dawson, T. P., Pearson, R. G. (2005). Selecting thresholds of ocurrence in the prediction of species distributions. *Ecography (Cop)* 28:385–393.

Liu, C., White, M., Newell, G. (2011). Measuring and comparing the accuracy of species distribution models with presence-absence data. *Ecography (Cop)* 34:232–243.

Mechoso, C. R., Arakawa, A. (2015). *Numerical Models: General Circulation Models,* Second Edi. Elsevier.

Oliveira-Filho, A. T. (2017). NeoTropTree, Flora arbórea da Região Neotropical: Um banco de dados envolvendo biogeografia, diversidade e conservação. [NeoTropTree, Neotropical Tree Flora: A database involving biogeography, diversity and conservation.] *Universidade Federal de Minas Gerais.* (http:// www.neotroptree,info).

Pacifici, M., Foden, W. B., Visconti, P., et al. (2015). Assessing species vulnerability to climate change. *Nat Clim Chang* 5:215–225.

Pearman, P. B., Guisan, A., Broennimann, O., Randin, C. F. (2008). Niche dynamics in space and time. *Trends Ecol Evol* 23:149–158.

Pearson, R. G. (2007). Species's distribution modelling for conservation educators and practitioners. *Lesson Conserv* 54–89.

Pearson, R. G., Dawson, T. P. (2003). Predicting the impacts of climate change on the distribution of species: Are bioclimate envelope models useful? *Glob Ecol Biogeogr* 12:361–371.

Peterson, A. T., Nakazawa, Y. (2008). Environmental data sets matter in ecological niche modelling: An example with Solenopsis invicta and Solenopsis richteri. *Glob Ecol Biogeogr* 17:135–144.

Peterson, A. T. (2003). Predicting the geography of species' invasions via ecological niche modeling. *Q Rev Biol* 78:419–33.

Peterson A. T., Soberon, J. (2012). Species Distribution Modeling and Ecological Niche Modeling: Getting the Concepts Right. *Nat Conserv* 10:102–107.

Concepts and Methods in Environmental Suitability Modeling ... 289

Peterson, A. T., Soberón, J., Pearson, R. G., et al. (2011). *Ecological niches and geographic distributions.*

Phillips, S. J., Anderson, R. P., Schapire, R. E. (2006). Maximum entropy modeling of species geographic distributions. *Ecol Modell* 190:231–259.

Quinn, G., Keough, M. (2002). *Experimental Design and Data Analysis for Biologists,* 2ª. Cambridge University Press, Cambridge.

R Core Team (2018). *R: A language and environment for statistical computing.*

Sillero, N., Carretero, M. A. (2013). Modelling the past and future distribution of contracting species. The Iberian lizard *Podarcis carbonelli* (Squamata: Lacertidae) as a case study. *Zool Anz* 252:289–298.

Siqueira, M. F. de, Durigan, G., de Marco-Júnior, P., Peterson, A. T. (2009). Something from nothing: Using landscape similarity and ecological niche modeling to find rare plant species. *J Nat Conserv* 17:25–32.

Slik, J. W. F., Arroyo-Rodríguez, V., Aiba, S-I., et al. (2015). An estimate of the number of tropical tree species. *Proc Natl Acad Sci* 112:7472–7477.

Soberón, J. (2007). Grinnellian and Eltonian niches and geographic distributions of species. *Ecol Lett* 10:1115–1123.

Soberon, J., Nakamura, M. (2009). Niches and distributional areas: Concepts, methods, and assumptions. *Proc Natl Acad Sci* 106:19644–19650.

Soberón, J., Peterson, A. T. (2005). Interpretation of models of fundamental ecological niches and species' distributional areas. *Biodivers Informatics* 2:1–10.

Thuiller, W., Georges, D., Engler, R., Breiner, F. (2016). *Ensemble Platform for Species Distribution Modeling.*

Thuiller, W., Lafourcade, B., Engler, R., Araújo, M. B. (2009). BIOMOD - A platform for ensemble forecasting of species distributions. *Ecography (Cop)* 32:369–373.

Troudet, J., Grandcolas, P., Blin, A., et al. (2017). Taxonomic bias in biodiversity data and societal preferences. *Sci Rep* 7:1–14.

van Zonneveld, M., Castañeda, N., Scheldeman, X, et al. (2014). Application of consensus theory to formalize expert evaluations of plant species distribution models. *Appl Veg Sci* 17:528–542.

Varela, S., Anderson, R. P., García-Valdés, R., Fernández-González, F. (2014). Environmental filters reduce the effects of sampling bias and improve predictions of ecological niche models. *Ecography (Cop)* 37:1084–1091.

Vaz, Ú. L., Nabout, J. C. (2016). Using ecological niche models to predict the impact of global climate change on the geographical distribution and productivity of *Euterpe oleracea* Mart. (Arecaceae) in the Amazon. *Acta Bot Brasilica* 30:1–6.

Vogler, R. E., Beltramino, A. A., Sede, M. M., et al. (2013). The giant *African snail*, Achatina fulica (Gastropoda: Achatinidae): Using bioclimatic models to identify South American areas susceptible to invasion. *Am Malacol Bull* 31:39–50.

Warren-Rhodes, K. A., Rhodes, K. L., Pointing, S. B., et al. (2006). Hypolithic cyanobacteria, dry limit of photosynthesis, and microbial ecology in the hyperarid Atacama Desert. *Microb Ecol* 52:389–398.

Willerslev, E., Hansen, A. J., Christensen, B., et al. (1999). Diversity of Holocene Life Forms in Fossil Glacier Ice. *Source Proc Natl Acad Sci United States Am* 96:8017–8021.

Wilson, E. (2000). *Sociobiology : the new synthesis.* Belknap Press of Harvard University Press.

Wilson, E. O. (2003). The encyclopedia of life. *Trends Ecol Evol* 18:77–80.

Wynn-Williams, D. D. (1996). Antarctic microbial diversity: The basis of polar ecosystem processes. *Biodivers Conserv* 5:1271–1293.

Zhang, L., Liu, S., Sun, P., et al. (2015). Consensus Forecasting of Species Distributions: The Effects of Niche Model Performance and Niche Properties. *PLoS One* 10:1–18.

Zwiener, V. P., Lira-Noriega, A., Grady, C. J., et al. (2017). Climate change as a driver of biotic homogenization of woody plants in the Atlantic Forest. *Glob Ecol Biogeogr* 27:298-309.

ABOUT THE EDITOR

Pedro V. Eisenlohr is an Assistant Professor at the State University of Mato Grosso (UNEMAT). He has more than 50 papers, most of them in reference journals such as PNAS, Biotropica, Diversity and Distributions, Ecological Modelling and Perspectives in Plant Ecology, Evolution and Systematics. He was Guest Editor of 'Biodiversity and Conservation' (2014-2015), Associate Editor of 'Biota Neotropica' (2015-2018) and is currently Associate Editor of 'Acta Botanica Brasilica'. He serves as a reviewer of journals such as Diversity & Distributions, Biotropica, Plant Ecology and Flora. He works mainly in the following subjects: vegetation ecology, ecological modeling, spatial autocorrelation, phytogeography, biogeographic transitions and biodiversity conservation.

INDEX

A

abundance, 4, 5, 9, 14, 16, 30, 34, 36, 38, 39, 105, 229, 230, 231

access, 55, 75, 118, 128, 210, 213, 214, 217

acetylcholinesterase, 55, 59

acetylcholinesterase inhibitor, 55, 59

ACL, 161, 162, 164

adaptability, 68, 167

adaptations, 47, 49, 169

agencies, x, 53, 212, 233, 256

agriculture, 20, 30, 44, 47, 51, 53, 60, 63, 75, 80, 84, 89, 92, 114, 117, 146, 148, 169, 173, 176, 178, 179, 180, 182, 195, 198, 199, 200, 201, 239, 240, 245, 246, 247, 259, 260, 261, 265

agroforestry biodiversity, 180

agroforestry systems, 174, 179, 182, 183, 185, 186, 187, 188, 193, 195, 202

Amazon, vi, ix, 10, 11, 36, 37, 70, 73, 74, 75, 86, 87, 97, 98, 100, 107, 119, 144, 148, 171, 172, 173, 175, 176, 177, 178, 179, 180, 181, 182, 183, 184, 187, 189, 190, 191, 193, 194, 196, 197, 198, 199, 200, 201, 202, 203, 204, 205, 225, 234, 238, 270, 291

Amazon River, 75, 87

amphibians, 69, 74, 76, 102, 107

Antarctic, 42, 43, 291

aptitude, 19, 20, 25, 30, 248

Areas For Permanent Preservation, vi, 114, 115, 117, 120, 121, 122, 125, 127, 128, 129, 130, 131, 133, 134, 135, 136, 137, 138, 140, 142

Argentina, ix, 41, 42, 43, 49, 50, 54, 56, 57, 58, 59, 60, 61, 63, 64, 66, 67, 75, 100, 154, 239, 240, 241, 243, 244, 245, 246, 251, 252, 257, 258, 259, 260, 261, 262, 263, 264, 265, 266, 267

arid deserts, 270

arid regions, 42, 43

armed conflict, 218, 219

assessment, 32, 33, 61, 68, 70, 103, 104, 200, 204, 205, 221, 224, 287, 288

atmosphere, 181, 182, 184, 185

Austria, 38, 101, 200, 205, 263

average cone weight, 161, 162

B

BAM diagram, 275, 276

basic needs, 67, 207

behaviors, 91

beneficiaries, 212

benefits, 56, 57, 92, 128, 133, 217, 218

bias, 15, 278, 279, 286, 291

biodiversity conservation, x, 1, 3, 10, 23, 25, 28, 30, 46, 48, 56, 88, 108, 114, 128, 148, 164, 172, 174, 220, 222, 223, 244, 285

biodiversity loss, 29, 87, 90, 172, 181, 183, 197

bioeconomy, 46, 52, 63

biological flow, 116, 118, 124, 125, 127, 134, 136, 137, 139, 142

biomass, 4, 6, 12, 34, 36, 179, 181, 184, 185, 186, 188, 189, 190, 191, 192, 195, 198, 199, 201, 204

biomes, 33, 43, 69, 72, 76, 77, 78, 82, 85, 93

bioprospecting, v, 41, 42, 46, 52, 53, 54, 56, 57, 58, 64

biotechnology, 52, 54, 56, 57, 244, 249, 258, 267

biotic, 5, 7, 35, 45, 48, 165, 184, 197, 202, 204, 275, 276, 286, 292

biotic interactions, 5, 35, 276, 286

Brazil, ix, 1, 9, 23, 36, 39, 40, 67, 70, 72, 75, 77, 79, 80, 82, 83, 84, 89, 92, 93, 94, 95, 96, 97, 98, 99, 100, 101, 103, 105, 107, 108, 110, 113, 118, 129, 143, 145, 146, 147, 150, 171, 172, 180, 187, 193, 197, 198, 205, 221, 225, 231, 234, 235, 252, 262, 269, 287

burn, 173, 178, 180, 185, 196, 198, 200

C

carbon, 99, 128, 164, 172, 174, 175, 176, 177, 179, 180, 181, 182, 183, 184, 185, 186, 187, 188, 190, 191, 194, 195, 196, 197, 198, 199, 200, 201, 202, 203, 204, 240, 284, 290

carbon emissions, 99, 182, 191, 284

carbon sequestration, 128, 164, 172, 202, 204, 240

carbon stocks, 175, 181, 182, 186, 194, 195, 198, 203

carbon storage, 172, 179, 182, 183, 188, 190, 204

Caribbean, 70, 96, 258

case study, 34, 35, 63, 145, 222, 287, 290

cattle, 20, 73, 75, 78, 80, 85, 245

CBD, 48, 50, 59, 256

CBNRM, 100

challenges, ix, 33, 34, 46, 57, 80, 92, 115, 141, 209, 219, 234, 241, 256, 261

CHELSA, 283, 284

chemical, 50, 54, 128, 230

civil society, xii, 68, 246

civilization, 177, 178

clarity, 117

classification, 3, 11, 119, 125, 145

climate change, 45, 46, 48, 52, 64, 68, 74, 81, 90, 92, 95, 99, 104, 106, 107, 108, 109, 111, 181, 183, 184, 195, 199, 202, 203, 204, 250, 271, 286, 288, 289, 291, 292

climates, 33, 245, 246, 287

CO_2 emissions (see also greenhouse gas emissions), 208

coastal communities, 247

Colombia, 9, 102

colonization, 124, 137, 178, 195, 216, 235, 247

commercial, 55, 82, 195, 226, 232, 242, 252

common species, 4, 6, 7, 9, 23, 27, 28, 31, 35, 37

communities, 2, 8, 34, 40, 53, 54, 56, 74, 77, 83, 97, 116, 118, 173, 177, 179, 180, 183, 188, 189, 195, 202, 204, 210, 211, 212, 214, 216, 217, 218, 219, 220, 221, 222, 226, 229, 230, 231, 232, 233, 235, 257, 260

Index

community, ix, 3, 4, 10, 32, 34, 40, 47, 57, 62, 68, 91, 98, 202, 211, 212, 217, 221, 223, 231, 232, 247

competition, 6, 38, 40, 79, 114, 165, 166, 271

complexity, 77, 114, 115, 176, 275, 280

compliance, 118, 211

composition, 16, 22, 38, 54, 97, 190, 198, 200, 201, 232, 240, 262

Condalia microphylla, 42, 45, 50, 51, 54, 58, 59, 64

cone dimension, 162

configuration, 137, 190

conflict of interest, 217

Congo, 270

Congress, iv, 58, 60, 89, 257, 258, 259, 261, 265, 266

congruence, 22, 23, 24, 26, 40, 204

connectivity, ix, 88, 92, 116, 117, 118, 123, 124, 125, 126, 134, 136, 137, 139, 140, 142, 143, 144, 149, 151, 185

consensus, 272, 283, 284, 291

conservation management, 233

conservation policies, 89, 209, 211, 219

conservation programs, 48, 52, 56, 216, 223

conservation strategies, 25, 29, 37, 39, 40, 48, 69, 87, 243, 244

conserving, 3, 26, 48, 50, 210, 217, 219

construction, 53, 72, 78, 83, 86, 96, 141, 234, 275, 278, 280, 282, 283

consumption, 46, 54, 80, 90, 100, 218, 226, 232

consumption patterns, 46

contamination, 51, 86, 97, 128

Convention on Biological Diversity, 50, 59

convergence, 184

coordination, 47, 117

correlation, 24, 26, 162, 255

correlation analysis, 26

correlation coefficient, 162

correlations, 157, 162

corruption, 217, 219

cosmetic, 50

cost, 70, 116, 139, 140, 150, 211, 217

cotton, 78

covering, 15, 120, 272

crocodilians, 72, 79, 81, 82, 85, 86, 95, 101, 105, 110

crop, 53, 80, 85, 168, 172, 176, 180, 196, 201, 203

crop irrigation, 85

crop production, 80

crops, 46, 59, 64, 76, 80, 173, 188, 192, 193, 194, 231, 260

crown, 168, 250

cryopreservation, 52, 242, 243, 259

cultivation, 80, 173, 176, 180, 183, 184, 185, 186, 187, 189, 191, 193, 195, 196, 205, 231, 244, 246

cultural heritage, 247

culture, 43, 48, 49, 51, 57, 58, 240, 241, 242, 247, 252, 254, 263, 265, 266

cycles, 172, 184, 186, 188, 191

D

data collection, 279

data processing, 131

database, 12, 23, 196, 278, 283, 284, 286, 287

death rate, 157

decision-making process, 223

decomposition, 183

decoration, 154

defaunation, 32, 90

defense mechanisms, 52

deforestation, 44, 47, 48, 49, 63, 68, 74, 75, 82, 84, 89, 90, 92, 98, 107, 117, 173, 175, 176, 177, 178, 179, 180, 181, 182, 183, 184, 185, 189, 191, 193, 194, 197, 198, 199, 200, 201, 202, 203, 204, 205, 208, 209, 212, 218, 220, 246

Index

degradation, 44, 48, 63, 78, 132, 136, 172, 173, 175, 180, 181, 182, 183, 184, 185, 191, 192, 204, 208, 209, 212, 213, 214, 215, 218, 219, 247, 254

degraded area, 46, 114, 115, 118, 128, 129, 130, 134, 141, 179, 188, 194, 241

degraded environments, 42

degraded forests, 195

dehydration, 63

Delta, 248, 262

democracy, 222

demography, 288

desertification, 42, 44, 45, 47, 63, 64, 245

destruction, 39, 43, 48, 72, 76, 79, 85, 100, 109

detachment, 232

detection, 145

developing countries, 209, 211, 212, 214, 216, 217, 218, 223, 232

diet, 86, 97, 226, 229, 231

discontinuity, 233

diseases, 92, 232, 254

disinfection, 51, 252

dispersion, 26, 123, 126, 127, 131, 133, 134, 137, 267, 271, 275

displacement, 123, 153, 218, 231, 276

distribution, 4, 5, 7, 8, 12, 27, 28, 35, 36, 39, 57, 69, 70, 76, 77, 82, 84, 86, 87, 88, 99, 104, 106, 107, 111, 130, 137, 270, 271, 273, 274, 275, 276, 280, 286, 287, 288, 289, 290, 291

disturbances, 43, 63, 81, 92, 137, 185, 191, 202, 230, 240

divergence, 38, 89

diversification, 76, 241

diversity, 32, 34, 37, 39, 40, 45, 48, 52, 54, 56, 69, 74, 76, 77, 83, 90, 91, 92, 94, 96, 109, 115, 136, 172, 184, 188, 189, 190, 196, 197, 198, 205, 230, 241, 242, 243, 244, 256, 260, 261, 269, 287, 291

diversity recovery, 190

DNA, 243

domestic demand, 189

domestication, 176, 201, 243

dominance, 51, 201

drought, 32, 42, 43, 190, 197, 229, 230, 253

drought periods, 42

drug discovery, 53

dry-land surface, 42

E

Easter, 198

ECM, 199

ecological connectivity, 185

ecological corridors, 116, 127, 135, 136, 138, 139, 142, 147, 149, 151

ecological processes, 78, 92, 132, 139, 142, 178, 184

ecological restoration, ix, 46, 53, 147, 148, 199, 249, 256

ecology, 32, 34, 35, 69, 91, 97, 100, 101, 103, 107, 108, 148, 149, 150, 197, 288, 291

ecosystem, 44, 45, 48, 49, 52, 53, 55, 56, 65, 66, 74, 89, 110, 115, 116, 128, 172, 173, 174, 175, 177, 178, 180, 181, 183, 184, 185, 188, 190, 195, 197, 204, 212, 213, 229, 233, 240, 241, 244, 250, 256, 263, 291

ecosystem services, 49, 55, 65, 74, 89, 110, 115, 181, 195, 204, 213, 233, 240, 241

Ecuador, 110, 180, 203

embryogenesis, 51, 58, 60, 243, 250, 253, 259

emergency, 246

emigration, 177, 178

emission, 180, 182, 183, 184, 185, 196, 212

endangered, 2, 32, 49, 50, 57, 59, 61, 62, 65, 70, 79, 83, 89, 111

endemic species, 22, 76

endemism, 11, 33, 37, 39, 76, 77, 78, 84, 104, 288

Index

endurance, 157, 165
energy, 45, 225
enforcement, 211, 220
engineering, 53
enlargement, 173
entropy, 104, 290
entropy model, 104, 290
environment, 16, 20, 38, 42, 45, 48, 58, 85, 86, 89, 102, 116, 117, 124, 167, 212, 215, 219, 228, 229, 230, 233, 254, 271, 274, 276, 279, 281, 290
environmental factors, 68, 229
environmental impact, 43, 56, 175, 188
environmental issues, 116, 233, 234
environmental layers, 271, 276, 277
environmental legislation, 89, 115, 117, 148
environmental management, 117, 221
environmental policy, 87, 90
environmental protection, 47, 243, 263
environmental quality, 178
environmental resources, 221
Environmental Rural Register, 117, 118, 120, 130
Environmental Suitability Modeling (ESM), vii, xix, 269, 270, 271, 281, 284, 285
environmental variables, 35, 270, 275, 280, 282, 284
environments, 29, 42, 43, 44, 47, 48, 49, 54, 56, 58, 63, 69, 70, 77, 80, 81, 83, 86, 88, 93, 110, 133, 225, 229, 230
evidence, 55, 104, 151, 173, 176, 177, 182, 184, 190, 194, 203, 221, 223
evolution, 40, 68, 77, 105, 109, 148, 222
examinations, 176
exploitation, 20, 76, 121, 238
extinction, 1, 20, 28, 29, 30, 31, 34, 37, 39, 40, 50, 68, 69, 70, 74, 79, 80, 81, 84, 87, 89, 90, 93, 95, 97, 98, 108, 114, 136, 241

F

families, 7, 68, 118, 180
farmers, 80, 187, 194
fauna, ix, 75, 91, 93, 99, 102, 116, 137, 139, 142, 151, 165, 225, 226, 230, 231, 232
fertility, 33, 45, 173, 178, 188, 192
fertilizers, 52
fibers, 50
field tests, 256
filters, 195, 279, 291
financial, 53, 121, 210
financial incentives, 53
financial resources, 210
fire, vi, ix, 12, 42, 43, 73, 80, 153, 154, 167, 169, 176, 184, 186, 193
fires, 80, 199
fish, 34, 226, 228, 229, 230, 231, 232, 233, 235
fitness, 69, 81
flexibility, 2, 69, 248, 285
flooding, 86, 87, 228, 229
flora, 8, 10, 11, 29, 31, 33, 35, 36, 37, 38, 47, 53, 56, 65, 77, 91, 93, 102, 105, 137, 139, 142, 204, 247, 262, 265
food, 45, 46, 50, 53, 54, 56, 57, 62, 67, 68, 82, 100, 172, 176, 178, 180, 183, 187, 194, 207, 215, 229, 231, 232, 233, 240, 241, 244, 248
food additive, 50
food industry, 54
food production, 46, 176
food security, 187, 194, 232, 240
force, 98
forest conservation, ix, 34, 47, 164, 182, 209, 210, 211, 215, 216, 217, 219, 265, 267
forest conservation planning, 164
forest ecosystem, ix, 46, 53, 56, 178, 181, 240, 242, 244, 248
forest fire, 47, 74, 240

forest fragmentation, 81, 83, 114, 197

forest fragments, 116, 133, 139, 144, 147, 189

forest genetic resources, 242, 263, 265, 266, 267

forest management, xix, 43, 91, 182, 186, 191, 195, 209, 211, 218, 221, 223, 240, 246

forest recovery, vi, ix, 113, 114, 115, 125, 135, 136, 138, 140, 146, 191, 202, 205

forest restoration, 115, 117, 118, 137, 146, 149, 150, 181, 183, 189, 262, 288

formation, 77, 85, 147, 210, 245, 248, 254

fragmentation, 44, 76, 78, 80, 83, 92, 98, 114, 116, 134, 137, 142, 145, 147, 148, 151, 175, 197, 208, 215, 233

fragmented landscapes, 114, 134, 147, 148

France, 9, 10, 153

fruits, 50, 53, 54, 60, 64, 229, 243, 255, 261

fuel woodcutting, 42

G

gene pool, 241

general circulation models (GCM), 283, 284, 289

genetic diversity, 52, 92, 242, 243, 244, 256

genetic resources, 47, 48, 49, 56, 64, 243, 249, 256, 259, 260, 264, 267

genus, 16, 39, 45, 51, 54, 82, 119, 226, 229, 279

Geoffroea decorticans, 42, 45, 50, 51, 55, 60, 64

geographic range, 2, 7, 8, 9, 12, 13, 14, 15, 16, 17, 33, 286

Geographical Information Systems, 114

geotechnologies, 120, 142, 150

Geotechnologies, 120

germination, 31, 45, 47, 50, 59, 62, 250, 251, 252

germplasm, 48, 49, 52, 61, 241, 244, 249, 256, 262

global climate change, 271, 291

global environment, 42, 87, 263

global scale, 184

global warming, 74, 81, 184, 197

goods and services, 44, 45, 66, 172, 173, 174, 175, 177, 178, 179, 180, 195, 240, 243

governance, 220, 222, 233

governments, 89, 209, 212

graph, 117, 123, 124, 134, 144, 151

grasslands, 12, 36, 38, 77, 247

greenhouse, 51, 90, 174, 184, 208, 254, 255

growth, 6, 30, 47, 51, 55, 59, 63, 80, 86, 90, 165, 166, 173, 177, 181, 182, 189, 198, 201, 215, 224, 243, 248, 250, 251, 252, 264

H

habitat, 2, 5, 8, 9, 11, 12, 13, 16, 18, 21, 22, 29, 37, 43, 44, 46, 48, 63, 68, 69, 72, 76, 79, 80, 82, 83, 84, 85, 86, 87, 88, 90, 91, 92, 95, 98, 109, 114, 115, 117, 123, 125, 137, 142, 145, 149, 155, 164, 169, 240, 253, 274, 288

habitat destruction, 39, 48, 72, 79, 109

habitat loss, 22, 29, 80, 82, 83, 84, 87, 88, 92, 102, 137

habitat quality, 86

habitat specificity, 8, 9, 16

habitat vulnerability, 9, 11, 18, 19, 21, 22

habitats, 8, 9, 10, 11, 15, 21, 27, 29, 30, 39, 43, 70, 74, 83, 91, 92, 116, 133, 248

height, 87, 157, 158, 159, 162, 165, 199

hermaphrodite, 255

herpetofauna, 69, 72, 74, 78, 88, 90, 91, 97, 98, 99

heterogeneity, 16, 77, 91, 137, 176

Index

history, 43, 69, 77, 82, 91, 94, 97, 117, 136, 140, 173, 176, 177, 190, 193, 196, 197, 203, 278
Holocene, 176, 203, 291
hotspots, 22, 26, 36, 37, 40, 68, 69, 72, 78, 88, 102, 103, 213
human, ix, 9, 20, 29, 31, 43, 45, 48, 49, 53, 54, 67, 74, 76, 77, 80, 84, 85, 87, 88, 90, 93, 97, 105, 107, 114, 173, 177, 178, 190, 195, 196, 197, 201, 202, 203, 204, 207, 215, 223, 233, 234, 240, 248
human activities, 20, 29, 49, 74, 80, 87, 93, 248
hunting, 68, 72, 76, 82, 83, 85, 95, 99, 105, 108, 180, 230, 231
hydroelectric dams, 72, 75, 78, 83, 85, 226, 230, 231

ICDPs, 212
identification, 92, 123, 241, 279
images, 114, 120, 145, 278
in vitro, 49, 50, 51, 52, 54, 59, 61, 63, 242, 252, 257, 258, 259, 261, 263, 265, 266
in vivo, 54
incidence, 214
income, 121, 187, 188, 194, 214, 218, 221, 223
income distribution, 214
income inequality, 214, 221
incubation period, 104
incubation time, 230
indexing, 49
India, 64, 221, 222, 223
individuals, 4, 8, 9, 14, 48, 55, 77, 85, 86, 116, 119, 123, 139, 189, 210, 231, 242, 256, 271
Indonesia, 196, 200, 222, 224
induction, 52, 253, 263
industries, 53, 252

industry, 244
ineffectiveness, 210
inequality, 214, 221
inflation, 280
infrastructure, 57, 75, 288
injuries, 101
insects, 110, 242, 248
institutions, 47, 222, 250
integration, 46, 47, 56, 57, 117, 179, 223
integrity, 233, 240
intellectual property, 55, 57
Intelligent Climate Agriculture, 174
interest groups, 278
interface, 286
interference, 232, 273
Intergovernmental Panel on Climate Change, 100, 281
Intergovernmental Panel On Climate Change, 100, 281
intervention, 114, 116, 117, 216
isolation, 77, 114, 134, 253
issues, 55, 57, 65, 167, 223, 232, 234, 272, 282
IUCN red list, 27, 28, 32, 68, 70, 100

juveniles, 85

L

lakes, 84, 91, 120
landscape connectivity, ix, 92, 117, 124, 125, 134, 136, 137, 139, 142, 144, 149, 151
landscape metrics, 116, 117, 139
landscape transformation, 114
landscapes, 50, 69, 72, 78, 80, 88, 94, 114, 115, 118, 134, 137, 142, 147, 148, 175, 183, 184, 190, 195, 197, 204

lead, 15, 77, 78, 114, 181, 184, 210, 213, 217, 218, 240, 244, 249, 256, 272, 274

legislation, 55, 89, 115, 117, 118, 127, 128, 131, 148

life cycle, 45, 133, 188

light, 3, 145, 165, 166

limestone, 78, 166

linnean shortfall, 28, 31

livestock, 44, 47, 51, 74, 80, 118, 173, 176, 187, 246

M

magnitude, 178

majority, 117, 128, 209, 213, 214

management, ix, 35, 43, 46, 48, 52, 55, 56, 57, 59, 66, 88, 91, 93, 96, 110, 114, 117, 131, 146, 148, 164, 173, 174, 175, 178, 179, 180, 181, 182, 183, 185, 186, 187, 188, 190, 194, 195, 200, 209, 210, 212, 220, 222, 223, 224, 233, 240, 251, 255, 256, 259, 260

mass, 31, 51, 74, 90, 159, 161

materials, 53, 55, 154, 167, 241, 247

matrix, 92, 139, 186, 189, 247, 282

medicine, 54, 67, 232, 235, 253, 254, 255

Mediterranean, 10, 38, 61, 94, 153, 154, 164, 167, 169

Mediterranean climate, 169

methodology, 249, 278

microhabitats, 87

micronutrients, 253

microorganisms, 53

migration, 137, 176, 178

misconceptions, 272

modelling, 286, 287, 288, 289

Monte, v, 41, 42, 43, 44, 48, 52, 53, 54, 55, 56, 58, 62, 63, 65, 66, 75, 103, 235, 243, 244, 245, 246, 248, 249, 250, 256, 262

Montenegro, 43, 44, 63

mortality, 6, 31, 82, 87, 105, 181, 197, 231, 251

multiplication, 46, 48, 49, 243, 248, 250

N

naphthalene, 51

native species, xii, xix, 42, 44, 48, 50, 52, 54, 58, 63, 64, 115, 140, 188, 241, 243, 244, 251, 254, 256, 257, 262

native tree species, ix, xix, 46, 241, 244, 249, 263

natural resource management, 212

natural resources, xiv, 49, 52, 78, 114, 115, 128, 141, 147, 175, 210, 211, 212, 217, 218, 234

negative consequences, 93, 183

neotropical, xii, xiii, 37, 40, 42, 43, 69, 70, 74, 75, 76, 97, 99, 104, 109, 190, 199, 203, 234, 235, 289

Nepal, 210, 212, 213, 222, 223

niche, xx, 5, 7, 33, 34, 38, 95, 270, 273, 274, 275, 276, 280, 286, 289, 290, 291, 292

niche limits, 5

nitrogen, 183

nodes, 123, 124, 125, 126

nucleation, 182, 189

nutrient, 49, 183, 225, 248

O

obstacles, 263

occurrence records, 274, 275, 278, 282

organism, 6, 123, 125, 274

Organization of American States, 196

organize, 53

organs, 86

overgrazing, 42, 44, 246

Index

301

P

Panama, 198

Paraguay, 75, 146, 245, 252

passive restoration, 179, 182, 186, 191, 194, 195

Patagonia, v, 41, 42, 43, 44, 46, 48, 49, 50, 54, 56, 58, 63, 64, 239, 245

Peru, 10, 97, 265

PES, 212, 213, 219

pests, 240, 242

pH, 5, 7, 37

phytoremediation, 258

pine trees, 154, 165, 166

Pinus halepensis, 167, 168, 169

plants, 34, 38, 40, 45, 49, 50, 51, 52, 55, 56, 58, 61, 62, 65, 66, 76, 89, 146, 169, 172, 203, 226, 228, 241, 242, 244, 250, 254, 257, 266, 292

PM, 38, 196, 197, 200, 201, 202, 203, 205

polar, 270, 291

policy, 29, 52, 56, 63, 69, 88, 89, 106, 141, 209, 211, 219, 222, 223, 257

policy makers, 88, 209, 219

pollination, 45, 137, 166, 241

polymorphism, 69

popular support, 216

population, 2, 6, 8, 14, 16, 29, 40, 45, 57, 65, 78, 79, 88, 90, 91, 93, 178, 195, 207, 210, 213, 215, 219, 224, 232, 233, 243, 244, 288

population growth, 88, 90, 178, 195, 210, 215, 219

population size, 2, 8, 14, 40, 78

Portugal, 153

poverty, 209, 213, 214, 215, 218, 219, 220, 223, 232

poverty alleviation, 232

poverty trap, 215, 220

precipitation, 7, 19, 21, 30, 43, 49, 119, 170, 190, 191, 192, 198, 229, 275, 283

predation, 34, 87, 230, 271

predictor variables, 278, 280, 282, 284, 285

preparation, iv, 54, 80, 154, 173, 231

preservation, ix, 55, 72, 114, 128, 131, 139, 142, 144, 145, 146, 147, 149, 150, 151, 193, 194, 242, 247, 264

Principal Components Analysis, 280

principles, 3, 34, 105, 257

prior knowledge, 271

priority areas, 36, 84, 87, 102, 134, 136, 137, 139

productivity, 45, 159, 165, 170, 174, 182, 183, 188, 189, 191, 192, 215, 240, 265, 291

project, 47, 64, 111, 170, 233, 246, 249, 270

propagation, ix, 42, 45, 46, 47, 49, 51, 52, 56, 57, 58, 59, 61, 62, 65, 184, 241, 242, 244, 248, 249, 250, 251, 254, 255, 256, 257, 258, 264, 265

Prosopis sp, 42, 50, 53, 62

protected areas, 50, 66, 72, 84, 87, 88, 93, 98, 99, 114, 139, 142, 149, 183, 187, 193, 194, 204, 211, 219, 221, 256

protection, 88, 92, 143, 144, 164, 233, 240, 247, 253

puna, 42, 43

Q

quality of life, 172, 173, 178, 179

quality of service, 46, 240

R

rain forest, 34, 110, 151, 196, 199

rainfall, 12, 20, 95, 155, 156, 230, 246

rainforest, 7, 10, 12, 31, 64, 75, 101, 109, 200, 203

rare species, 2, 3, 4, 6, 7, 8, 10, 11, 14, 15, 22, 23, 24, 25, 26, 27, 28, 29, 30, 31, 33, 35, 36, 39, 40, 278 rate of change, 208

recovery, ix, 46, 48, 58, 76, 91, 114, 115, 116, 120, 125, 131, 135, 136, 138, 140, 141, 142, 144, 146, 179, 181, 185, 186, 187, 189, 190, 191, 194, 195, 202, 205, 233, 249

Red List, 36, 68, 70, 72, 100

REDD+, vi, ix, 171, 175, 178, 180, 181, 182, 183, 185, 186, 189, 191, 195, 196, 200, 205, 212

reforestation, 42, 44, 51, 154, 164, 198

regeneration, 46, 51, 65, 179, 182, 184, 187, 188, 193, 194, 195, 198, 203, 243, 252

regions of the world, 11, 49

regression, 157, 162, 286

regression analysis, 157

regrowth, 196, 199

regulations, 194, 211, 216

regulatory agencies, 210

regulatory system, 209

rehabilitation, 179, 181, 182, 186, 194

relevance, 100, 117, 118, 124, 132, 140, 142, 256, 285

remote sensing, 144, 148, 198

Representative Concentration Pathways (RCP), 283

reproduction, 44, 45, 87, 102, 139, 230, 233, 248

reptile species, 68, 69, 70, 79, 91

reptiles, v, ix, 67, 68, 69, 70, 74, 76, 79, 81, 82, 87, 91, 92, 93, 94, 95, 97, 98, 100, 102, 104, 237

requirements, 6, 15, 29, 45, 46, 114, 136, 139, 140, 141, 193, 270

research institutions, 57

researchers, ix, 178, 209, 233, 271, 272

resilience, 45, 184, 185, 191, 199, 200, 203, 205

resource management, 48, 219, 221

respiratory problems, 232

response, 7, 32, 35, 49, 51, 137, 184, 209, 215, 246, 255, 271, 279, 280

restoration, vi, 29, 42, 46, 47, 50, 52, 53, 64, 65, 92, 115, 116, 117, 118, 120, 121, 123, 124, 125, 127, 130, 131, 134, 136, 137, 139, 140, 141, 145, 146, 147, 148, 149, 150, 174, 179, 181, 182, 183, 186, 187, 189, 191, 194, 195, 199, 205, 239, 241, 244, 248, 249, 256, 260, 262, 263, 266, 288

restoration programs, 42, 46

risk, 30, 34, 68, 69, 70, 72, 87, 89, 93, 97, 98, 101, 108, 183, 230, 259, 270

roots, 51, 52, 166, 250, 252, 253, 255

rotations, 193

routines, 121, 281, 284

rules, 210, 211, 212

rural population, 56, 207, 213

rural poverty, 214, 215, 217

S

safety, 64

salinity, 5, 7

samplings, 249

sanctuaries, 92

Sartorius, 94

SAS, 157

satellite monitoring, 89

science, 40, 56, 62, 92, 204, 221, 222, 264, 271, 272, 288

scientific knowledge, 47

secondary forests, 173, 180, 182, 184, 186, 187, 188, 189, 190, 191, 193, 194, 199, 201, 202, 203

seed, 8, 34, 40, 47, 49, 51, 62, 137, 146, 154, 157, 159, 161, 162, 163, 164, 166, 167, 168, 184, 189, 193, 241, 242, 243, 249, 250, 255, 256, 260, 262

seedlings, 47, 50, 51, 64, 128, 165, 241, 243, 252, 256

semi-arid, 42, 43, 45, 47, 48, 50, 51, 56, 63, 65, 154, 245

Index   303

semi-arid environments, 42, 43, 47

semidesert regions, 42

services, iv, 42, 45, 49, 53, 55, 65, 74, 89, 110, 115, 174, 181, 195, 204, 212, 213, 233, 240, 241

settlements, 117, 143, 148, 173, 177, 178, 180, 193, 195

sex, 81, 82, 95, 97, 101, 104, 105, 107, 109, 229

sex ratio, 81, 95, 101, 104, 105, 107, 229

shifting cultivation, 173, 176, 180, 183, 184, 185, 186, 187, 189, 191, 193, 195, 205

showing, 4, 24, 76, 84, 156, 162, 233

shrub desert, 42, 43

society, ix, 42, 46, 68, 115, 240, 263

software, 120, 121, 122, 123, 125, 126, 127, 149

South America, 33, 36, 60, 65, 68, 70, 75, 95, 99, 107, 109, 172, 203, 246, 249, 251, 291

spatial distribution patterns, 7

species distribution, 5, 7, 35, 37, 84, 88, 111, 271, 273, 274, 275, 276, 286, 287, 288, 289, 290, 291, 292

species richness, 23, 26, 29, 33, 34, 37, 68, 83, 90, 91, 104, 185, 190, 200, 240, 249, 256, 288

stability, 45, 116, 181

stakeholders, 154, 234

state, 12, 17, 73, 75, 76, 85, 97, 98, 115, 117, 132, 146, 148, 185, 198, 210, 212, 228

storage, 48, 49, 52, 172, 179, 182, 183, 188, 190, 204, 242, 243

stress, 5, 45, 49, 52, 66, 165, 248

structural characteristics, 100, 262

structure, 34, 83, 95, 105, 114, 116, 147, 151, 168, 173, 178, 179, 183, 190, 195, 230, 231, 240, 248, 262, 265, 267, 286

subsistence, 83, 176, 180, 231

succession, 182, 183, 186, 187, 189, 190, 191, 193, 194, 195, 197, 198, 199, 202

surface area, 15, 166, 288

survival, 45, 47, 64, 77, 81, 87, 92, 97, 114, 116, 139, 156, 251, 255, 273

sustainability, 43, 141, 151, 178, 240

sustainable agriculture, 240

sustainable development, 55, 56, 59, 167, 193, 202, 211, 232, 233

sustainable forest management, 240

sustainable management, ix, 46, 59, 93, 173, 175, 180, 181, 182, 186, 187, 194, 195, 209, 240

sustainable production, 56, 114

Switzerland, 10, 31, 95, 100, 101, 143, 200

synchronization, 187

synthesis, ix, 54, 98, 287, 291

T

tannins, 253

target, 115, 211, 249, 256, 262

taxa, 7, 28, 35, 37, 79, 104

taxonomy, 36

teams, 233

techniques, 46, 47, 48, 49, 115, 116, 242, 243, 244, 246, 250, 270, 278

technological change, 44

technological developments, 56

temperature, 7, 81, 82, 86, 87, 104, 107, 155, 156, 165, 229, 230, 275, 283

term plans, 115

terrestrial ecosystems, 45, 68, 204

territorial, 115, 121, 147

threatened species, 27, 49, 68, 70, 72, 88, 93, 287

tissue, 48, 49, 52, 58, 63, 243, 264, 266

tobacco, 63, 264

traits, 134, 139, 156, 159, 170

transformation, 43, 85, 114, 176, 181, 243

treatment, 55, 251, 254

tree species, v, vi, ix, 1, 10, 12, 13, 15, 18, 22, 23, 24, 27, 29, 31, 36, 40, 46, 50,

154, 156, 172, 186, 190, 239, 241, 243, 244, 247, 249, 253, 256, 263, 267, 290

Tropical Africa, vi, 207, 215

tropical forests, 143, 172, 176, 184, 190, 196, 197, 201, 270, 285

U

uniform, 90

United Nations Convention to Combat Desertification, 44

United States, 6, 101, 169, 291

universe, 271, 276

V

validation, 65, 118

valorization, 53

variables, 7, 23, 37, 115, 120, 124, 134, 157, 162, 168, 185, 270, 275, 276, 277, 279, 280, 281, 282, 283, 284, 285

vegetation, 11, 33, 36, 38, 42, 43, 45, 62, 76, 77, 80, 84, 89, 91, 98, 114, 115, 119, 120, 124, 125, 127, 128, 129, 130, 132, 136, 137, 143, 146, 155, 176, 183, 185, 193, 195, 199, 229, 262, 265, 282

vegetative propagation, 47, 52, 58, 242, 255, 257, 258, 264

velocity, 230

vertebrates, 69, 76, 83, 109

vessels, 231

Vietnam, 63

viral pathogens, 242

vulnerability, 2, 3, 9, 11, 18, 19, 20, 21, 22, 30, 40, 68, 95, 104, 111, 234, 289

vulnerability principles, 3

W

Washington, 63, 99, 109, 169, 223, 235

water, 43, 45, 49, 52, 66, 80, 81, 85, 86, 91, 120, 128, 130, 139, 144, 164, 170, 190, 207, 225, 229, 230, 233, 240, 253

water quality, 139

watershed, 151, 222

web, 66, 278, 284

welfare, 217

well-being, 174, 233

wetlands, 92

wilderness, 36

wildlife, 49, 50, 65, 76, 80, 91, 92, 108, 111, 143, 164, 183, 220, 222, 223, 233

wildlife conservation, 143

wood, 12, 20, 47, 50, 154, 155, 164, 242, 247, 252, 255

wood species, 242

woodland, 36, 247

woody species, 46, 47, 48, 50, 51, 62, 205, 243

woody vegetation, 38, 42, 43, 45, 262, 265

World Bank, 53, 214, 215, 216, 223

WorldClim, 21

X

xerophyte, 247

Z

Zimbabwe, 214

Geomatics and Conservation Biology

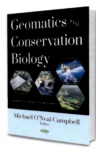

Editor: Michael O'Neal Campbell (Camosun College, Victoria, BC, Canada)

Series: Conservation Biology and Biodiversity

Book Description: This edited book, composed of chapters written by scholars of the geomatics-based, environmental and biological sciences, examines selected topics from the intersecting fields of geomatics (including remote sensing, geographical information science (GIS), global positioning systems (GPS), mapping and field survey methods) and conservation biology (including ecology and conservation policy), with case studies from West Africa, Canada, India and Malaysia.

Hardcover ISBN: 978-1-53614-468-0
Retail Price: $230

Forest Ecosystems: Management, Impact Assessment and Conservation

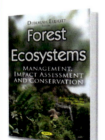

Editor: Deborah Elliott

Series: Wildlife Protection, Destruction and Extinction

Book Description: This current book reviews and analyzes forest ecosystems. Chapter One begins with a discussion of radioactivity in forest ecosystems. Chapter Two discusses how litter chemistry has significant effects on soil biogeochemistry and looks into the relationships between litter chemistry, soil chemistry and microbial activity.

Hardcover ISBN: 978-1-63485-794-9
Retail Price: $95

Old-Growth Forests and Coniferous Forests: Ecology, Habitat and Conservation

Editor: Ronald P. Weber

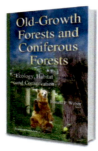

Series: Environmental Research Advances

Book Description: This book discusses the ecology, habitat and conservation of old-growth forests, as well as coniferous forests.

Hardcover ISBN: 978-1-63482-369-2
Retail Price: $130

Forest Ecosystems: Biodiversity, Management and Conservation

Editor: Noel C. Roberts

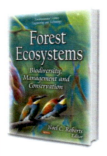

Series: Environmental Science, Engineering and Technology

Book Description: This book discusses the management and conservation of forest ecosystems in several areas in the world which include the Neotropics, Norway, United Kingdom and Siberia.

Hardcover ISBN: 978-1-63117-815-3
Retail Price: $130